HISTORIC WRITINGS ON HYDRAULICS

HISTORIC WRITINGS ON HYDRAULICS

A Catalogue of the History of Hydraulics Collection in the University of Iowa Libraries

Iowa City

FRIENDS OF THE UNIVERSITY OF IOWA LIBRARIES

1984

Library of Congress Cataloging in Publication Data

Rouse, Hunter, 1906–
Historic writings on hydraulics.

Sequel to: Hydraulics in the United States 1776–1976.
Includes index.
1. Hydraulics—History—Bibliography—Catalogs.
2. Hydraulic engineering—History—Bibliography—
Catalogs. 3. University of Iowa. Libraries—Catalogs.
I. Hitchcock, Leonard. II. Title.
Z5853.H9R59 1984 [TC145] 016.6201'06 84–10195
ISBN 0-87414–031–5

Printed in the United States of America

This catalogue,
which Hunter Rouse and the late Frank S. Hanlin, Bibliographer,
The University of Iowa Libraries,
had long planned to develop as a joint enterprise, is now published
and dedicated to the memory of

FRANK S. HANLIN

1924–1982

PREFACE

As recounted in the Preface of *History of Hydraulics* (see item 453), the senior author of that book, who is also the writer of this Preface, had the privilege of spending the winter of 1952–53 in the library of the Etablishments Neyrpic at Grenoble, France. The primary attraction there was the wealth of historical material on hydraulics which had been gathered by his host, Pierre Danel. Upon his return to Iowa City, the writer set out to assemble a similar collection of source material. He used for this purpose the income from the *History* as well as from other publications of the Iowa Institute of Hydraulic Research, of which he was then director. During the ten or twelve years following publication of the *History*, some 400 relatively rare books were thus acquired.

In every instance, location of the books and negotiations with the booksellers were conducted by the late Frank Hanlin, bibliographer of the University of Iowa Libraries. In some cases Institute funds were matched by the University, and in a few others the University met the entire cost. A very recent addition resulted from a private gift of 150 items by the late Dr. F. T. Mavis, formerly head of Mechanics and Hydraulics at Iowa; about half of these were originally in the possession of Sherman M. Woodward, Mavis's predecessor at Iowa; many of them were either new to the collection or in better condition than those already at hand. Counting the individual books in multi-volume works, over 500 volumes are now contained in the History of Hydraulics Collection at Iowa.

The term "hydraulics" has, of course, many different connotations, and such a collection could logically emphasize one or more different phases of endeavor. As noted in the *History*, "Of primary interest to practicing engineers would be the matter of improvement of hydraulic works. Laboratory investigators would focus attention upon the evolution of experimental techniques. Of importance to manufacturing concerns would be the development of hydraulic machinery. And the general historian would consider more essential the relationship of progress in the one field to that of science as a whole. The authors of the *History* have chosen as their primary theme the formulation of the underlying principles of fluid motion, incorporating such aspects of experimentation, engineering application, related scientific progress, and world history as seemed to bear upon that theme." The collection herein reviewed consistently reflects the same point of view.

The principal items in this collection, although by no means all, are discussed in the *History*, which also reviews the lives of many of their authors in some detail. Note should likewise be taken of the *History*'s sequel, *Hydraulics in the United States 1776–1976* (item 458), which encompasses many American authors not included in its forerunner. In the following pages less will be said about the authors themselves, and their works will be commented upon only briefly. For more extensive coverage, reference is made to corresponding portions of the two histories and to the various historical and encyclopedic works cited therein.

Instead of following the more customary alphabetical or strict chronological arrangement of the individual entries, these have been introduced in order of initial publication date of each author. Successive editions of the same work are then listed in sequence, and other writings by the same author follow directly according to date (for example, items 80–84, 137–144, 196–198). In several cases, booklets represent memoirs of leaders' lives or appraisals of their accomplishments; these complete the listings of their publications. All titles have been kept in their original language without translation, but descriptive matter contained on the title pages has often been rendered into English in the subsequent commentary.

In most cases the items are original works, first editions when possible, later ones when the first were not to be had. In a few instances photocopies have been included, the originals not being available and the works being considered too important to omit. Journal papers are listed only when reprinted as separates. On the other hand, the several volumes of collected works by leaders in the field consist almost entirely of reprinted papers, books by these authors being available separately. Noteworthy in another regard are the frequent *Raccolta* or reprinted writings of Italian hydraulicians (items 108, 179, 271), some appearing not once or twice but even more often.

It is believed that the History of Hydraulics Collection reviewed herein includes most of the important works covered by the writer's definition of the subject, not to mention many of lesser merit presented for the sake of completeness. Others that emphasize hydraulic engineering are included for their illustration of the current state of the art or science. Except for more recent historical reviews, coverage has arbitrarily been terminated with the first quarter of the present century. In glancing through the review, one will soon note that the concentration of activity shifts more or less successively from country to country; i.e., from Greece to Rome, and later from Italy to France, to England, to Germany, and eventually to the United States, the Soviet Union not receiving due emphasis only for lack of better communication with the West. Somewhat the same situation exists with regard to early Arabic manuscripts which were so important to the transfer of knowledge from Greek to Latin, for much of it would otherwise have been lost.

Selection and arrangement of the items in this catalogue and tentative preparation of the commentaries were undertaken by the writer during a month of residence at Iowa City as guest of the Iowa Institute of Hydraulic Research under Dr. John F. Kennedy, presently its director. After another month of editing by the writer at his retirement home in Arizona, compilation of the bibliographic data was carried out for the most part by Leonard Hitchcock, while preparation of the manuscript for the printer was made by Rijn Templeton under the direction of Frank Paluka, head of the Special Collections Department in the University of Iowa Libraries. This publication was financed by the Institute of Hydraulic Research and by the Friends of the University of Iowa Libraries, as well as by a bequest from the estate of Edith L. Kelly in memory of William Galt Raymond, first Dean of Engineering at The University of Iowa, and his wife.

November 1983

Hunter Rouse
Carver Professor and
Engineering Dean Emeritus

A Catalogue of the History of Hydraulics Collection in the University of Iowa Libraries

1
Aristotle (384–322 B.C.)

S. THOMAE AQVINATIS COMMENTARIA IN OCTO PHYSICORUM ARISTOTELIS LIBROS . . . Ad haec accessit Roberti Linconiensis in eosdem Summa. [Edited by Romulus Fabius] Venetiis [Venice]: Apud Ivntas, 1566.

[vi], 167 leaves; illus.; 32 x 22 cm.

Aristotle was an encyclopedist, whose writings ranged from physics to metaphysics. Though he taught that knowledge must continually advance, his teachings (physics as well as metaphysics) eventually came to be crystallized—particularly by the Scholastics under St. Thomas Aquinas (1225?–1274).

2
Archimedes (287–212 B.C.)

DE IIS QVAE VEHVNTVR IN ACQVA LIBRI DVO, a Federico Commandino in pristinvm nitorem resticvti et commentariis illvstrati. Bononiae [Bologna]: Ex officina Alexandri Benacii, 1565.

[iv] 43 (i.e. 45), [iv] 47 leaves; illus.; 21 x 15 cm. Bound with the author's Liber de centro gravitatis solidorvm.

Generally considered the foundation of hydrostatics, Archimedes' principles were actually restricted to buoyancy and flotation.

3
Archimedes (287–212 B.C.)

DE INSIDENTIBVS AQVAE. Liber primvs [and] Liber secvndvs. [Edited by Niccolò Tartaglia] Venetiis [Venice]: Apud Cvrtium Troianvm, 1565.

6, [i] 16 leaves; illus.; 20 x 14 cm.

4
Archimedes (287–212 B.C.)

WORKS . . . With a supplement, The method of Archimedes, recently discovered by Heiberg. Edited in modern notation with introductory chapters by T.L. Heath. New York: Dover Publications, [195–?]

clxxxvi, 326, 51 pp.; illus.; 20 x 13 cm.

5
Philo, of Byzantium (fl. ca. 250 B.C.)

LE LIVRE DES APPAREILS PNEUMATIQUES ET DES MACHINES HYDRAULIQUES. . . . Édité d'après les versions arabes d'Oxford et de Constantinople et traduit en Français, par le baron Carra de Vaux. Paris: C. Klincksieck, 1903.

[27]-235 pp.; illus.; 27 x 20 cm. (Notices et extraits des manuscrits de la Bibliothèque Nationale et autres bibliothèques publiés. L'Académie des Inscriptions et Belles-Lettres. Paris: Imprimerie Nationale, vol. 38.)

Elementary pneumatics and application of the siphon principle to many ingenious systems.

6
HERO, of Alexandria (fl. 62 A.D.)
SPIRITALIVM LIBER. A Federico Commandino ex graeco, nvper in latinum conversvs. Urbini [Urbino] 1575.
70 (i.e. 80) pp.; illus.; 22 x 15 cm.

7
HERO, of Alexandria (fl. 62 A.D.)
DE GLI AVTOMATI, OVERO MACHINE SE MOVENTI . . . tradotti dal Greco da Bernardino Baldi . . . Nouamente ristampato, con ogni diligenza riccoretto. . . . In Venetia [Venice]: Appresso Gio. Battista Bertoni, libraro al Pellegri.
47 leaves; illus.; 20 x 14 cm. Engraved title page.

Encyclopedic treatment of pneumatics, mechanics, water clocks, and magical tricks, including early suggestions of the continuity principle.

8
HERO, of Alexandria (fl. 62 A.D.)
BUCH VON LUFFT-UND WASSER-KUNSTEN. . . . [Translated from Greek into Latin by Friderich Commandino von Urbin with an appendix by Joannis Baptistae Aleotti] [Translated from Latin into German by Agathum Carionem] Franckfurt: Johann Wilhelm Ammon, 1688.
[viii] 158, 32 pp.; illus.; frontis.; 20 x 15 cm.

9
HERO, of Alexandria (fl 62 A.D.)
THE PNEUMATICS OF HERO OF ALEXANDRIA, from the original Greek. Translated for and edited by Bennet Woodcroft. London: Charles Whittingham, 1851.
xix, 117, 2 pp.; illus.; 24 x 16 cm.

10
Rochas d'Aiglun, Albert de (1837–1914)
LA SCIENCE DES PHILOSOPHES ET L'ART DES THAUMATURGES DANS L'ANTIQUITÉ. Paris: G. Masson, 1882

218, [ii] pp.; xxiv leaves of plates; 25 x 16 cm. Includes: Les pneumatiques de Héron and Fragment des pneumatiques de Philon. Les pneumatiques de Héron d'Alexandrie traduite pour la première fois du Grec en Français par M. A. de Rochas, 15 pp., is included separately in paper wrapper.

11
Rochas, d'Aliglun, Albert de (1837–1914)
LA SCIENCE DES PHILOSOPHES ET L'ART DES THAUMATURGES DANS L'ANTIQUITÉ. Seconde édition augmentée de documents inédits et accompagnée de 24 planches. Paris: Dorbon-ainé,. 192–?.

250, [2] pp.; 24 pp. of plates; 24 x 16 cm. Includes: Les pneumatiques de Héron d'Alexandrie; Notes relatives au texte Grec des pneumatiques de Héron d'Alexandrie; Fragment des pneumatiques de Philon de Bysance; Mécaniques de Héron. . . . L'optique et la catoprique d'Euclide. . . . Extraits des philosophoumena d'origène.

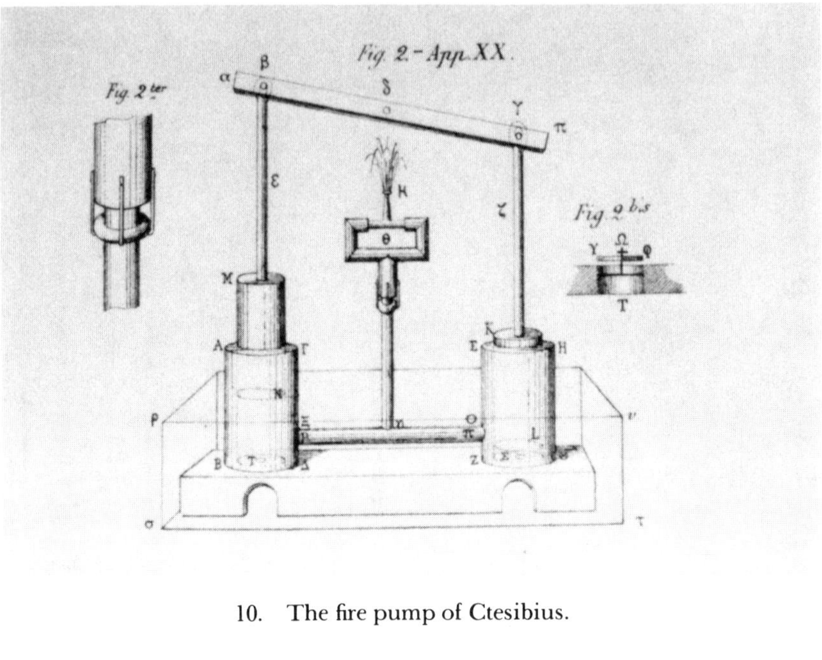

10. The fire pump of Ctesibius.

12
Vitruvius Pollio (1st century B.C.)
VITRVVIVS ITERVM ET FRONTINVS A IOCVNDO REVISI REPVRGATIQVE QVAN TVM EX COLLATIONE LICVIT. [Florentiae] [Florence]: [Sumptibus.P. de Giunta, 1513]
[viii] pp., 187, 24, [24] leaves; illus.; 16 x 10 cm.

The ten books of this treatise covered all contemporary aspects of architecture, civil works, and military construction, three of them dealing with water.

12. Undershot water wheel after Vitruvius.

13
VITRUVIUS POLLIO (1st century B.C.)
VITRUVIUS: THE TEN BOOKS ON ARCHTECTURE. Translated by Morris Hicky Morgan. With illustrations and original designs prepared under the direction of Herbert Longford Warren. New York: Dover Publications, [1960]
xii, 331 pp. [+16 pp. adv.]; illus.; 20 x 14 cm.

14
FRONTINUS, SEXTUS JULIUS (40–104 A.D.)
DE AQVAE DVCTIBVS VRBIS ROMAE COMMENTARIVS. Antiqua fidei restitutus, atque explicatus opera et stvdio Ioannis Poleni. Patavii [Padua], 1722.
[viii] 32, 396 pp.; illus.; 15 leaves of plates; 25 x 19 cm.

A complete description is given herein of the Roman methods of water distribution, with annotation by Giovanni Poleni.

15
FRONTINUS, SEXTUS JULIUS (40–104 A.D.)
THE TWO BOOKS ON THE WATER SUPPLY OF THE CITY OF ROME OF SEXTUS JULIUS FRONTINUS . . . A photographic reproduction of the sole original Latin manuscript, and its reprint in Latin; also a translation into English, and explanatory chapters, by Clemens Herschel. Boston: Dana Estes, 1899.
xxiv, [ii] 296 pp.; illus.; frontis.; 23 pp., 3 leaves (all folding) of plates; 29 x 21 cm.

This English version contains not only a translation from the Latin but also an extensive commentary by Clemens Herschel.

16
FRONTINUS, SEXTUS JULIUS (40–104 A.D.)
THE TWO BOOKS ON THE WATER SUPPLY OF THE CITY OF ROME OF SEXTUS JULIUS FRONTINUS . . . A photographic reproduction of the sole original Latin manuscript, and its reprint in Latin; also a translation into English, and explanatory chapters, by Clemens Herschel. Second edition. New York: Longmans, Green and Co., 1913.
xxiv. [ii] 296 pp.; illus.; frontis.; 23 pp., 3 leaves (all folding) of plates; 29 x 21 cm.

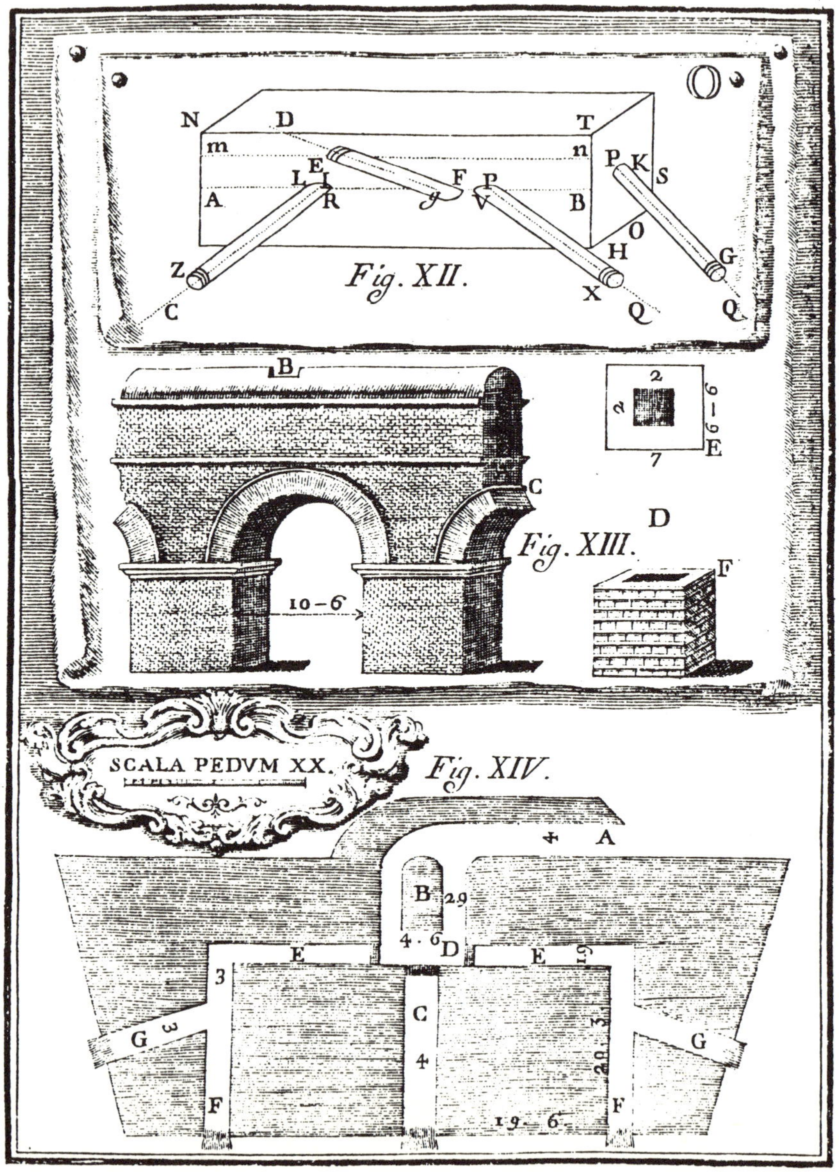

14. Aqueduct details, after Frontinus.

17
al-Jazarī, Ismā'īl ibn al-Razzāz (fl. 1181–1204)
THE BOOK OF KNOWLEDGE OF INGENIOUS MECHANICAL DEVICES. Translated [from the Arabic] and annotated by Donald R. Hill. Dordrecht, Holland: D. Reidel, 1974.
xxv, 285 pp.; illus.; frontis.; 34 x 25 cm. Number 176 of a limited edition.

18
Kyeser, Conrad (*b.* 1366)
BELLIFORTIS. Düsseldorf: VDI-Verlag, 1967.
Vol. 1: 284 pp. (facsim., some color); 33 x 25 cm.
Vol. 2: liv, 107 pp.; illus.; 33 x 25 cm. Half-title: Umschrift und Übersetzung von Götz Quarg.

19
Mariano di Jacopo (1381–*ca.*1458)
[MARIANO TACCOLA] DE MACHINIS; the engineering treatise of 1449. Introduction, Latin texts, descriptions of engines and technical commentaries by Gustina Scaglia. Wiesbaden: Dr. Ludwig Reichert, 1971.
Vol. 1: 181 pp.; 30 x 23 cm.
Vol. 2: 210 pp. of plates (facsim.); 30 x 23 cm.

20
Leonardo da Vinci (1452–1519)
DEL MOTO E MISURA DELL'ACQUA. Bologna: Francesco Cardinali, 1828. [viii], 273–450 pp.; 51 leaves of plates; 25 x 19 cm.

Many portions of Leonardo's notebooks dealing with water (including the first correct statement of the continuity principle) have been gathered together and the original drawings copied.

21
Leonardo da Vinci (1452–1519)
DEL MOTO E MISURA DELL'ACQUA . . . editi sul codice archetipo barberiniano a cura di E. Carusi ed A. Favaro. Bologna: Nicola Zanichelli, [1923]
xxiii, 408, [3] pp.; illus.; frontis.; 4 pp., 3 leaves of plates; 24 x 16 cm. Publication of the Istituto di Studii Vinciani in Roma. Nuova serie, Testi Vinciani, vol. 1.

22
LEONARDO DA VINCI (1452–1519)
ÉTUDES ET DESSINS SUR LES CANAUX. . . . Paris: Edouard Rouveyre, 1901.
[30] leaves; [14] leaves of plates (facsim., 11 folding); 37 x 28 cm.

23
LEONARDO DA VINCI (1452–1519)
I LIBRI DEL VOLO DI LEONARDO DA VINCI. Nella ricostruzione critica di Arturo Uccelli con la collaborazione di Carlo Zammattio. Milano: Ulrico Hoepli, [1952]
xxxiv, 233, [2] pp.; illus.; 27 x 22 cm.

24
MOSTRA DELLA SCIENZA E TECNICA DI LEONARDO. Milano: Museo Nazionale
Della Scienza E Della Tecnica, 1953–[1954?]
Vol. 1: 82, [4] pp.; 1 leaf, 6 pp. of plates; 21 x 15 cm.
Vol. 2: 40 pp.; 24 pp. of plates; 21 x 15 cm.
Vol. 3: 45 pp.; 21 x 15 cm.
Vol. 4: 14 pp.; 1 leaf, 12 pp. of plates; 21 x 15 cm.
Vol. 5: 37, [1] pp.; 1 leaf (folding), 16 pp. of plates; 21 x 15 cm.
Vol. 6: 45, [1] pp.; 7 leaves of plates: 21 x 15 cm.
Vol. 7: 98, [1] pp.; illus.; 21 x 15 cm.
Vol. 8: 59, [1] pp.; [2] leaves; illus.; 21 x 15 cm.
Vol. 9: 101, [1] pp.; illus.; 21 x 15 cm.

25
DUHEM, PIERRE (*b.* 1861)
ÉTUDES SUR LÉONARD DE VINCI. Vols 1 and 2, Ceux qu'il a lus et ceux qui l'ont lu; Vol. 3, Les précurseurs parisiens de Galilée. Paris: F. de Nobele, 1955.
Vol. 1: vii, 355 pp.; illus.; 22 x 14 cm.
Vol. 2: iv, 474 pp.; illus.; 22 x 14 cm.
Vol. 3: xiv, 605 pp.; illus.; 22 x 14 cm.

A scholarly analysis of those whom Leonardo read and those who read Leonardo ("Ceux qu'il a lus et ceux qui l'ont lu").

26
BELLINCIONI, GIOVANNI
LEONARDO DA VINCI, FONDATORE DELLA SCIENZA IDRAULICA. Firenze [Florence]: C.O.S.P.I.T., [1952]
130 pp.; illus.; 17 x 11 cm.

27
LEONARDO DA VINCI (1452–1519)
THE NOTEBOOKS OF LEONARDO DA VINCI. Arranged, rendered into English and introduced by Edward MacCurdy. Garden City, New York: Garden City Publishing Co., 1941.
1247 pp.; frontis.; 63 leaves of plates; 23 x 15 cm.

28
TARTAGLIA, NICOLO (*d.* 1557)
RAGIONAMENTI DE NICOLO TARTAGLIA SOPRA LA SUA TRAVAGLIATA INVENTIONE. . . . [Venetia] [Venice]: [Stampata . . . per Nicolo Bascarini 1551]
[48] pp.; illus.; 19 x 14 cm.

29
CORNARO, LUIGI (1475–1566)
TRATTATO DI ACQVE DEL MAGNIFICO. In Padoua [Padua]: Appresso Gratioso Perchacino, 1560.
30 [i.e. 60] pp.; 1 leaf of plates (double); 21 x 15 cm.

30
[CEREDI, GIUSEPPI] (*fl.* 1565)
TRE DISCORSI SOPRA IL MODO D'ALZAR ACQVE DA'LVOGHI BASSI. . . . In Parma: Appresso Seth Viotti, 1567.
[xviii] 100 pp.; illus.; 3 leaves of plates (all folding); 20 x 16 cm.

31
AIMO, BATTISTA (16th c.)
TRACTATVS DE FLVVIORVM ALLVVIONIBVS, deq́; ijs, quae ex aluuione nascuntur commodis, & incommodis. . . . Venetiis [Venice]: Apud Francisicum Zilettum, 1581.
[xx] 307, [23] pp.; illus.; 2 leaves of plates (both folding); 30 x 21 cm.

32
AIMO, BATTISTA (16th c.)
DE ALLUUIONIBUS; TRACTATVS FERTILLISSIMVS. In quo incrementi & decrementi fluminum, ad fundorum vicinorum commoda & incommoda, cum naturali acquisitione & amissione . . . tractatur. . . . Venetiis [Venice]: Apud Baretium Baretium, & Socios, 1599.
[xx] 307, [23] pp.; illus.; 2 leaves of plates (both double); 29 x 20 cm.

33
AIMO, BATTISTA (16th c.)
DE ALLUVIONIBUS, TRACTATVS FERTILISSIMVS. [Lipsiae] [Leipzig]: Prodiit sumtus in typographeio suo faciente H. Grosio, 1601.
[xx] 307, [22] pp.; illus.; 2 leaves of plates (both double); 32 x 20 cm.

34
RAMELLI, AGOSTINO (1531–ca. 1600)
LE DIVERSE ET ARTIFICIOSE MACHINE, DEL CAPITANO AGOSTINO RAMELLI . . . Nellequali si contengono uarij et industriosi Mouimenti, degni digrandissima Speculatione, per cauarne beneficio infinito in ogni forte d'operatione. A Parigi: in casa del' autore, 1588.
[xxx], 338 pp.; illus.; 33 x 23 cm.

35
RAMELLI, AGOSTINO (1531–ca. 1600)
THE VARIOUS AND INGENIOUS MACHINES OF AGOSTINO RAMELLI (1588). Translated from the Italian and French with a biographical study of the author by Martha Teach Gnudi. Technical annotations and pictorial glossary by Eugene S. Ferguson. Baltimore: Johns Hopkins University Press, 1976.
604 pp.; illus.; frontis.; 35 x 23 cm.

36
CARAZZI, CARLO
TRATTATO DELL'ALLVVIONI DI TVTTII FIVMI ET TORRENTI. . . . Bologna: Presso gli Heredi di Giouanni Rossi, 1602.
[vi] 199, 23, [+4] pp.; illus.; 10 leaves of plates (1 folding, 4 double); 29 x 20 cm.

37
[CASTELLI, BENEDETTO] (1577–1644)
RIPOSTA ALLE OPPOSIZIONI DEL S. LODOVICO DELLE COLOMBE E DEL S. VINCENZIO DI GRAZIA, contro al trattato del sig. Galileo Galilei, delle cose che stanno sù l'acqua, ò che in quella si muouono. In Firenze [Florence]: Appresso Cosimo Giunti, 1615.
[iv], 319 (i.e. 335) [+5] pp.; 20 x 14 cm.

Contrary to Leonardo, Galileo had many pupils, the most productive of whom was Castelli.

38
Castelli, Benedetto (1577–1644)
DELLA MISVRA DELL'ACQVE CORRENTI. Roma: Nella stamparia Camerale, 1628.
[iv] 59 pp. [+i p. errata]; illus.; 20 x 15 cm. Engraved title page. Bound with the author's Della misvra dell'acqve correnti, 1660, and several works by other authors.

This work contains the rediscovery of the continuity principle, apparently lost with Leonardo's notebooks.

39
Castelli, Benedetto (1577–1644)
DELLA MISVRA DELL'ACQVE CORRENTI. Bologna: H.H. del Dozza, 1660.
184 pp.; illus.; 20 x 15 cm.

40
Fludd, Robert (1574–1637)
UTRIUSQUE COSMI MAIORIS SCILICET ET MINORIS METAPHYSICA, PHYSICA ATQVE TECHNICA HISTORIA in dua volumina secundum cosmi differentiam diuisa. Oppenhemii [Oppenheim]: aere Johan-Theodori de Bry, typis Hieronymi Galleri, Vol. 1, 1617; Vol. 2, 1618
[ii] 206 [6], [ii] 788 [10] pp.; illus.; 31 x 19 cm. With engraved title pages for both volumes. Two volumes in one.

41
Stevin, Simon (1548–1620)
NIEVVVE MANIERE VAN STERCTEBOV, DOOR SPILSLUYSEN. Rotterdam: Ian van Waesberghe, 1617.
[iv] 59 [+2] pp.; illus.; 30 x 19 cm.

A Dutch mathematician and military engineer, Stevin extended the principles of hydrostatics by evaluating the pressure of water on boundary surfaces and thereby explaining the hydrostatic paradox.

42
Stevin, Simon (1548–1620)
LES OEUVRES MATHÉMATIQUES DE SIMON STEVIN DE BRUGES. Reveu, corrigé, & augmenté par Albert Girard. A Leyde: Bonaventure & Abráham Elsevier, 1634.
[viii] 222 [+2], 678 pp.; illus.; 34 x 22 cm.

43
Stevin, Simon (1548–1620)
THE PRINCIPAL WORKS OF SIMON STEVIN. Volume I: General introduction; mechanics. Edited by E. J. Dijksterhuis, [translated into English by C. Dikshoorn] Amsterdam: C.V. Swets & Zeitlinger, 1955.
[i] ii–v, 617 pp.; illus.; frontis.; 24 x 17 cm.

44
Fontana, Giovanni (1540–1614)
MESURE RACCOLTE . . . dell'accrescimento che hanno fatto li fiumi, torrenti, e fossi che hanno causato l'inondatione à Roma il Natale, 1598. In Roma: Appresso Antonio Maria Gioiosi, 1640.
[vi] 18 pp. Photocopy of the original accompanied by a translation into English, produced by the Engineering Societies Library, N.Y., 1964.

45
Mersenne, Marin (1588–1648)
COGITATA PHYSICO MATHEMATICA. In quibus tam naturae quàm artis effectus admirandi certissimis demonstrationibus explicantur. Parisiis [Paris]: Sumptibus Antonii Bertier, 1644.
[xxxvi] 40, [xx] 11–14 [+2] 41–214, [viii] 225–370, [xvi] 48, 47–48, 49–96, [xii] 140 pp. [+20 pp. index]; illus.; 23 x 17 cm.

46
Mersenne, Marin (1588–1648)
NOVARVM OBSERVATIONVM PHYSICO-MATHEMATICARVM. Parisiis [Paris]: Sumptibus Antonii Bertier, 1647.
[xxxii] 62 [+2] 63–235 [+3] 4 pp.; illus.; 23 x 18 cm.

A supplement to Mersenne's Cogitata Physico-Mathematica, 1644, (item 45) and his Universae Geometriae Mixtaeque Mathematicae Synopsis, 1644.

47
Torricelli, Evangelista (1608–1647)
OPERA GEOMETRICA. [Florentiae [Florence]: Typis Amatoris Masse & Laurentij de Landis, 1644].
[iv] 3–243, [ii] 3–56, 55–78, 81–136, 129–136, 145–115 (i.e., 151) pp.; illus.; 21 x 15 cm.

Though not described in this work, it was Torricelli who applied his mentor's (not Galileo's) analysis of free fall to the trajectory of a liquid jet and who made the first water barometer.

48
FRANÇOIS, JEAN (1582–1668)
LA SCIENCE DES EAVX qvi expliqve en qvatre parties levr formation, commvnication, movvemens, et meslanges. Avec les arts de conduire les eavx, et mesvrer la grandevr tant des eavx qve des terres. . . . A Paris: Florentin Lambert, [1655].
[xvi] 120, 40, 24, 56, 28, 68 pp.; illus.; 20 x 14 cm.

49
GALILEI, GALILEO (1564–1642)
DISCORSO AL SERENISSIMO DON COSIMO II . . . intorno alle cose, che stanno sù l'acqua, ò che in quella si muouono. Seconda editione. In Bologna: Per gli H.H. del Dozza, 1655.
[iv] 68 pp.; illus.; 22 x 15 cm.

Galileo supplemented Aristotle's practice of reasoning and Leonardo's of observing by his own of experimenting, nevertheless concluding that he understood more about astronomy than about hydraulics.

50
GALILEI, GALILEO (1564–1642)
OPERE. [Edited by Carlo Manolessi] Bologna: Per gli H.H. del Dozza, 1656.
Vol. 1: [xxii] 43, [ii] 3–48, [viii] 160, [iv] 68, [ii] 3–127, [iv] 264, [ii] 3–43 pp.; illus.; frontis.; 2 leaves of plates (1 folding); 23 x 16 cm.
Vol. 2: [iv] 3–60, [ii] 3–7 [+1], 105 [+3] 105–156, [ii] 3–48, [viii] 179 [+1], [ii] 53–126, [viii] 238 [+6] pp: illus.; 23 x 16 cm.

51
GALILEI, GALILEO (1564–1642)
DIALOGUES CONCERNING TWO NEW SCIENCES. Translated from the Italian and Latin into English by Henry Crew and Alfonso de Salvio, with an introduction by Antonio Favaro. Evanston and Chicago: Northwestern University, 1939.
[v] vi–xxi, [iv] 300 pp.; illus.; frontis.; 22 x 18 cm.

52
GALILEI, GALILEO (1564–1642)
A DISCOURSE . . . CONCERNING THE NATATION OF BODIES UPON, AND SUBMERSION IN, THE WATER. Englished from the second edition of the Italian . . . by Thomas Salusbury, Esq. London: Printed by William Leybourn, 1663.
[ii] 76 pp.; illus.; 28 x 21 cm. Facsimile edition, under title Discourse

on Bodies in Water, with introduction and notes by Stillman Drake, published by University of Illinois Press, Urbana, Ill., 1960. (Facsimile Reprints in the history of science, sponsored by the History of Science Society of the University of Illinois, No. 2)

53
ARREDI, FILIPPO (*b.* 1905)
L'IDRAULICA DI GALILEO E DELLA SUA SCUOLA. Roma: Scuola Tipografica Pio X, [1942].
38 pp.; 24 x 17 cm. Paper from a conference held at the R. Università di Firenze, March 10, 1942.

54
SALUSBURY, THOMAS (*d.* 1666.)
MATHEMATICAL COLLECTIONS AND TRANSLATIONS: IN TWO TOMES. London: Printed by William Leybourn, Vol. 1, 1661; Vol. 2; 1665.
Vol. 1: [xvi] 503 [+25], [xiv] 118 [+5] pp.; illus.; 32 x 21 cm.
Vol. 2: [iv] 469 [+1], [iv] 483–516 [517] pp.; illus.; 32 x 21 cm.
Facsimile edition, with an analytical and bio-bibliographical introduction by Stillman Drake, published by Dawsons of Pall Mall (London) and Zeitlin and VerBrugge (Los Angeles) in 1967.

Veritable tomes containing works by Galileo, Castelli, Descartes, Archimedes, Tartaglia and others.

55
BARATTIERE, GIOVANNI BATTISTA (1601–1677)
ARCHITETTURA D'ACQVE. . . . Piacenza: Gio. Bazachi, 1656.
[xx] 277 [3] pp.; illus.; 1 leaf of plate (folding); 29 x 19 cm.

56
BOYLE, ROBERT (1627–1691)
NEW EXPERIMENTS PHYSICO-MECHANICAL, TOUCHING THE AIR . . . Whereunto is added a defense of the author's explication of the experiments, against the objections of Franciscus Linus, and, Thomas Hobbes. The second edition. [Oxford: Printed by H. Hall . . . for Tho. Robinson, 1662 (New experiments physico-mechanical. . . .)] [London: Printed by F.G. for Thomas Robinson, 1662 (A defense of the doctrine . . . against . . . Linus; An examen of Mr. T. Hobbes his Dialogus physicus de naturâ aëris.)]
[xvi] 207, [xii] 122, [viii] 86, 85–98 pp.; 2 leaves of plates (1 folding); 19 x 15 cm. Contents (each with own title page): [1] New experiments

physico-mechanical touching the spring of the air, . . . [2] A defense of the doctrine touching the spring and weight of the air . . . against the objections of Franciscus Linus. . . . [3] An examen of Mr. T. Hobbes his Dialogus physicus de naturâ aëris. . . .

57
Boyle, Robert (1627–1691)
HYDROSTATICAL PARADOXES, MADE OUT BY NEW EXPERIMENTS, (for the most part physical and easie.) Oxford: Printed by William Hall for Richard Davis, 1666.
[xxxvi], 245 pp.; illus.; 3 leaves of plates (all folding); 17 x 10 cm.

58
Boyle, Robert (1627–1691)
TRACTS WRITTEN BY THE HONORABLE ROBERT BOYLE, containing: New experiments touching the relation betwixt flame and air. And about explosions; An hydrostatical discourse occasion'd by some objections of Dr. Henry More. . . . London: Printed for Richard Davis, 1672.
[vi], 142 [ii], 17, [xii], 176, [ii], 19, [ii], 16, [ii], 39, [+1] pp.; 17 x 11 cm.

59
Boyle, Robert (1627–1691)
PARADOXA HYDROSTATICA NOVIS EXPERIMENTIS (MAXIMAM PARTEM PHYSICIS AC FACILIBUS) EVICTA, et Regiae Britanniarvm Societati exhibita. Genevae [Geneva]: Apud Samuelem de Tournes, 1680.
[xvi], 72 pp.; 3 leaves of plates (all folding); 21 x 17 cm.

60
Boyle, Robert (1627–1691)
EXERCITATIONES DE ATMOSPHAERIS CORPORUM CONSISTENTIUM; déque mira subtilitate, determinata natura, & insigni vi effluviorum. Subjunctis experimentis novis, ostendentibus, posse partes ignis & flammae reddi stabiles ponderabilésque. Unà cum detecta penetrabilitate vitri à ponderabilibus partibus flammae. Genevae [Geneva]: Apud Samuelem de Tournes, 1680.
[ii], 12, 55, 21, [2], 9 pp.; 21 x 17 cm.

61
Dobrzensky, Jacobus Joannes Wenceslaus (*fl.* 1659–*ca.*1680)
NOVA' ET AMAENIOR DE ADMIRANDO FONTIVM GENIO (EX

ABDITIS NATVRAE CLAVSTRIS, IN ORBIS LVCEM EMANANTE) PHILOSOPHIA. Ferrariae [Ferrara]: Apud Alphonsum, 1659.

[xxiv] 121 [2] pp.; illus.; 31 x 21 cm. Added, engraved title page, dated 1657.

62
BONINI, FILIPPO MARIA

IL TEVERE INCATENATO OVERO L'ARTE DI FRENAR L'ACQUE CORRENTI. In Roma: Stampa di Francesco Moneta, 1663.

[xvi], 422 pp. [2, 20 pp. index]; illus.; 25 x 18 cm.

63
PASCAL, BLAISE (1623–1662)

TRAITEZ DE L'EQVILIBRE DES LIQVEVRS ET DE LA PESANTEVR DE LA MASSE DE L'AIR. Contenant l'explication des causes de divers effets de la nature qui n'avoient point esté bien connus jusques ici, & particulieremét de ceux que l'on avoit attribuez à l'horreur du vuide. A Paris: Gvilluame Desprez, 1663.

[xxviii] 232 pp. [+8 pp. table]; illus.; 2 leaves of plates (all folding); 14 x 8 cm.

Herein are the completing statements of hydrostatics that liquid pressure is transmitted equally in all directions, that air has weight, and that the barometric pressure is affected by elevation and by atmospheric change.

64
PASCAL, BLAISE (1623–1662)

TRAITEZ DE L'EQVILIBRE DES LIQVEVRS, ET DE LA PESANTEVR DE LA MASSE DE L'AIR. . . . Second édition. A Paris: Gvillaume Desprez, 1664.

[xxvi], 232 pp. [+6 pp. table]; illus.; 2 leaves of plates (all folding); 15 x 8 cm.

65
PASCAL, BLAISE (1623–1662)

TRAITEZ DE L'EQULIBRE DES LIQUEURS, ET DE LA PESANTEUR DE LA MASSE DE L'AIR. . . . A Paris: Guillaume Desprez, 1698.

[xxvi], 238 pp.; illus.; 2 leaves of plates (all folding); 16 x 9 cm.

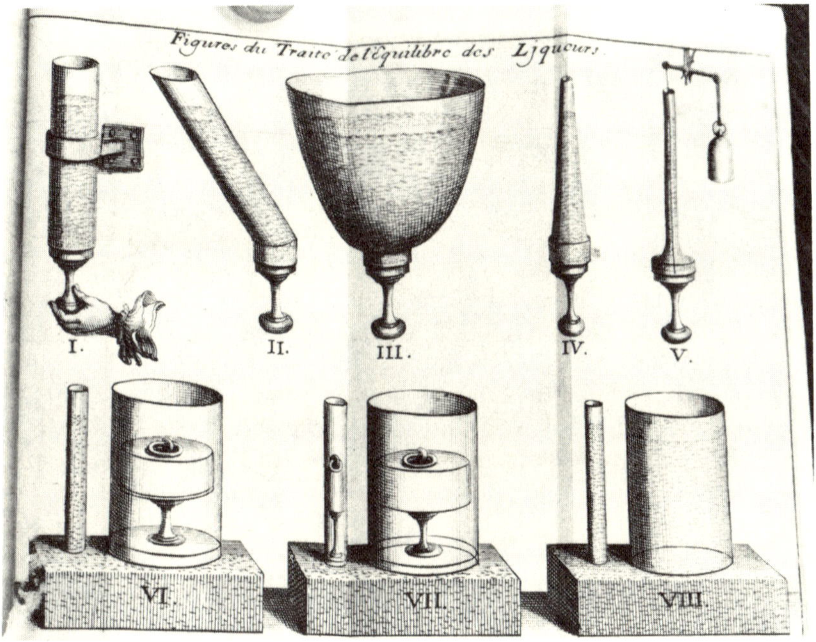

63. Pascal's treatment of hydrostatic pressure.

66
PASCAL, BLAISE (1632–1662)
THE PHYSICAL TREATISES OF PASCAL: The equilibrium of liquids and the weight of the mass of the air. Translated by I.H.B. and A.G.H. Spiers, with introduction and notes by Frederick Barry. New York: Columbia University Press, 1937.
[v] vi–xxviii, 181 pp. [+2 pp. adv.]; illus.; 3 leaves of plates (all folding); 23 x 15 cm. (Records of civilization; sources and studies, no. XXVIII)

67
SCHOTT, GASPAR (1608–1666)
ANATOMIA PHYSICO-HYDROSTATICA FONTIUM AC FLUMINUM, LIBRIS VI, EXPLICATA. . . . Herbipoli [Wūrzburg]: Sumptibus viduae Jo. Godefridi . . . excudit Jobvs Hertz, 1663.
[xxii], 433 pp. [+14 pp. index, 1 p. errata]; 14 leaves of plates (1 folding); 16 x 9 cm. Engraved added title-page.

68
MICHELINI, FAMIANO (1592?–1666)
TRATTATO DELLA DIREZIONE DE'FIVMI, nel quale si dimostrano da' suoi veri principi i modi più sicuri, e meno dispendiosi di riparare a'danni, che sogliono farsi dall'acque. Firenze: Stamperia della Stella, 1664.
[xvi], 151 [1] pp.; 9 leaves of plates (all folding); 26 x 18 cm.

69
GUERICKE, OTTO VON (1602–1686)
EXPERIMENTA NOVA (UT VOCANTUR) MAGDEBURGICA DE VACUO SPATIO. Primùm à R.P. Gaspare Schotto . . . nunc verò ab ipso auctore perfectiùs edita . . . aucta. Amstelodami [Amsterdam]: Apud Joannem Janssonium à Waesberge, 1672.
[xiv], 244 pp. [+4 pp. index]; illus.; 3 leaves of plates (2 folding); 33 x 20 cm. Two of the original plates are missing and have been replaced with photocopies.

70
[HALE, SIR MATTHEW (1609–1676)]
DIFFICILES NVGAE: or, observations touching the Torricellian experiment, and the various solutions of the same, especially touching the weight and elasticity of the air. The second edition, with some occasional additions. London: Printed by W. Godbid, for William Shrowsbury, 1675.
[vi], 304 pp. [+5 pp. table, 2 pp. adv.], 48 pp.; frontis.; 2 leaves of plates (both folding); 17 x 11 cm. Bound with the author's Observations touching the principles of natural motions (item 70a).

70a
[HALE, SIR MATTHEW (1609–1676)]
OBSERVATIONS TOUCHING THE PRINCIPLES OF NATURAL MOTIONS; and especially touching rarefaction & condensation: together with a reply to certain remarks touching the gravitation of fluids. London: Printed by W. Godbid, for W. Shrowsbury, 1677.
[xiv], 285 pp. [3 pp. adv.]; illus.; 17 x 11 cm. Bound with the author's Difficiles nvgae. . . . (item 70).

71
FABRETTI, RAFFAELE (1618–1700)
DE AQVIS ET AQVAEDVCTIBVS VETERIS ROMAE DISSERTA-

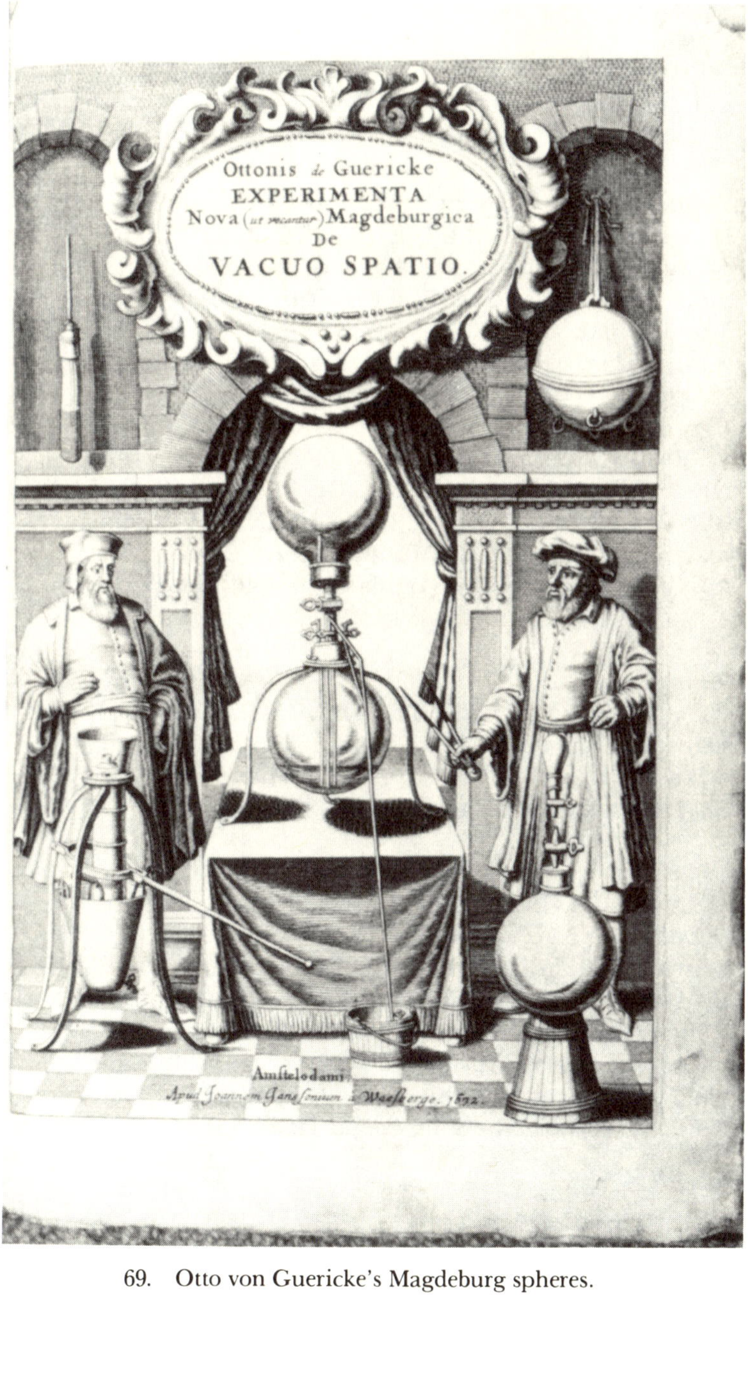

69. Otto von Guericke's Magdeburg spheres.

TIONES TRES. Romae [Rome]: Typis Ioannis Baptistae Busotti, 1680.

[ii] 3–193 pp. [+22 pp. index]; illus.; 3 leaves of plates (all folding); 24 x 17 cm.

72
[Chérubin d'Orléans, le pere (1613–1697)]

L'EXPERIENCE JUSTIFIEE, POUR L'ELEVATION DES EAUX, par un nouveau moyen à tel hauteur, & en telle quantité que ce soit, proportionnant la force. Par monsieur L.C.D.O. A Paris: François Muguet, 1681.

[xiv], 52 pp. [+1 p. errata]; 14 x 8 cm.

73
Ceva, Giovanni (17th c.)

OPUSCULA MATHEMATICA DE POTENTIIS OBLIQVIS, DE PENDVLIS, DE VASIS, ET DE FLVMINIBVS. Mediolani [Milan]: Ex typographia Ludouici Montiae, 1682.

[iv], 53 [+2] pp.; 8 leaves of plates (all folding); 25 x 19 cm.

74
Bernoulli, Jacques (1654–1705)

DISSERTATIO DE GRAVITATE AETHERIS. Amstelaedami [Amsterdam]: Apud Henr. Wetstenium, 1683.

[xiv] 269 pp. [+3 pp. index]; 4 leaves of plates (all folding); 16 x 9 cm. Engraved added title-page.

75
ESSAYES OF NATURAL EXPERIMENTS MADE IN THE ACADEMIE DEL CIMENTO . . .

written in Italian by the Secretary of that academy. Translated by Richard Waller. With a new introduction [for the facsimile edition] by A. Rupert Hall. New York: Johnson Reprint Corporation, 1964.

[vii] viii–xx, [xxiii], 160 pp. [+10 pp. table]; frontis.; 19 leaves of plates; 23 x 15 cm. Facsimile of the 1684 edition, London, Printed for Benjamin Alsop. (The Sources of Science).

76
Capra, Domenico (17th c.)

IL VERO RIPARO, IL FACILE, IL NATVRALE PER OUUIARE, Ō RIMEDIARE OGNI CORROSIONE, E RUINA DI FIUME, E

TORRENTE, ABBENCHE GIUDICATA IRREMEDIABILE. In Bologna: Giacomo Monti, 1685.
[xii], 72 pp.; 7 leaves of plates (all folding); 20 x 14 cm.

77
MEIJER, CORNELIS (*b. ca.* 1640)
L'ARTE DI RESTITUIRE À ROMA LA TRALASCIATA NAUIGATIONE DEL SUO TEUERE. . . . In Roma: Stamperia del Lazzari Varese, 1685.
[viii] [162] 15 pp.; illus.; 5 leaves of plates (2 double); 41 x 27 cm. Engraved added title-page.

78
BORELLI, GIOVANNI ALFONSO (1608–1679)
DE MOTIONIBUS NATURALIBUS, A GRAVITATE PENDENTIBUS. Lugduni Batavorum [Leyden, Netherlands]: Apud Petrum Vander Aa, 1686.
[iv], 360 [+25 pp. indices, 7 pp. adv.]; 14 leaves of plates (all folding); 20 x 14 cm.

79
BÖCKLER, GEORG ANDREAS (*fl.* 1648–1685)
THEATRUM MACHINARUM NOVUM, exhibens aquarias, alatas, jumentarias . . . molas. Variis frumentis commolendis . . . adaptas: cum utilibus & rarioribus hydrotechnematis, . . . Ex Germaniâ in Latium recens translatum operâ R.D. Henrici Schmitz. Noribergae [Nuremberg]: Apud Pauli Fürsti, 1686.
[xii], 55 pp.; 154 leaves of plates; 134 x 21 cm. Engraved title page dated 1662. Bound with the author's Architectura curiosa nova (item 79a).

Profusely illustrated treatise on contemporary machinery.

79a
BÖCKLER, GEORG ANDREAS (*fl.* 1648–1685)
ARCHITECTURA CURIOSA NOVA, EXPONENS 1. Fundamenta hydragogica, indolemq; . . . 2. Varios aquarum ac salientium fontium. . . . 3. Magnum amoenissimorum fontium, machinarumq; . . . 4. Specus artificiales sumtuosissimas, cum plerisq; . . . 5. Cum auctario figurarum elegantissimarum, ad hortorum topiaria vario ductu dividenda, nec non conclavium laquearibus ac pavimentis segmentandis, itemq;. . . . Et in Latinam linguam translata à Johanne

Christophoro Sturmio. Norimbergae [Nuremberg]: Impensis Pauli Fürsten, [n.d.].

[viii], 30, [ii], 13, [ii] 22, [ii], 26 pp.; 200 leaves of plates; 134 x 21 cm. Engraved added title page. Bound with the author's Theatrum machinarum novum . . . (item 79).

80
MARIOTTE, EDME (1620?–1684)

TRAITE DU MOUVEMENT DES EAUX ET DES AUTRES CORPS FLUIDES. . . . Mis en lumiere par les soins de M. De La Hire. A Paris: Estienne Michallet, 1686.

[xiv], 408 pp. [+14 pp. table, 6 pp. errata]; illus.; 16 x 9 cm.

Prior of a monastery near Dijon, one of the founders of the Académie Royale des Sciences, and considered the father of the French school of experimental hydraulics, Mariotte contributed to knowledge of jet mechanics, wind pressure, pipe resistance, flow measurement, and in particular the elasticity of the air.

81
MARIOTTE, EDME (1620?–1684)

TRAITÉ DU MOUVEMENT DES EAUX ET DES AUTRES CORPS FLUIDES. Nouvelle édition corrigée. Mis en lumiere par les soins de M. De La Hire. A Paris: Jean Jombert, 1700.

[xii], 390 pp. [+14 pp. table, 4 pp. adv.]; illus.; 16 x 9 cm.

82
MARIOTTE, EDME (1620?–1684)

TRAITE DU MOUVEMENT DES EAUX ET DES AUTRES CORPS FLUIDES. Nouvelle édition corrigée & augmentée des regles pour les jets d'eau. Mis en lumiere par les soins de M.De La Hire. A Paris: Claude-Jombert, 1718.

[xii], 414 pp. [+13 pp. table, 5 pp. adv.]; illus.; 1 leaf of plates (folding); 16 x 9 cm.

83
MARIOTTE, EDME (1620?–1684)

THE MOTION OF WATER, AND OTHER FLUIDS. Being a treatise of hydrostaticks. . . . together with a little treatise . . . giving practical rules for fountains, or jets d'eau. [Translated by J.T. Desaguliers] by whom are added, several annotations for explaining the doubtful places. London: Printed for J. Senex and W. Taylor, 1718.

[ii] iii–xxiii [+i p. errata], 290 pp. [+2 pp. adv.]; 7 leaves of plates (all folding); 19 x 12 cm.

84
MARIOTTE, EDME (1620?–1684)
OEUVRES DE M. MARIOTTE, DE L'ACADÉMIE ROYALE DES SCIENCES; comprenant tous les traitez de cet auteur, tant ceux qui avoient déja paru séparément, que ceux qui n'avoient pas encore eté publiés. . . . Nouvelle édition. A La Haye: Jean Neaulme, 1740.
Vol. 1: [xii] 3–320 pp.; 13 leaves of plates (all folding); 25 x 19 cm.
Vol. 2: [iv] [321]–701 pp. [33 pp. contents, 2 pp. adv.]; 13 leaves of plates (all folding); 25 x 19 cm.

85
BIOT, JEAN BAPTISTE (1774–1862)
[ARTICLES ON THE] MEMOIRS OF THE LIFE, WRITINGS AND DISCOVERIES OF SIR ISAAC NEWTON . . . [by] Sir David Brewster. [Paris]: [Imprimerie Impériale] [1855].
32 pp.; 27 x 22 cm. Extracted from the Journal des Savants, cahiers d'octobre et de novembre, 1855.

86
MACLAURIN, COLIN (1698–1746)
AN ACCOUNT OF SIR ISAAC NEWTON'S PHILISOPHICAL DISCOVERIES, IN FOUR BOOKS. Second edition. Published from the author's manuscript papers by Patrick Murdoch. London: Printed for A. Millar, 1750.
[x] xxvi, 412 pp.; 5 leaves of plates (all folding); 20 x 12 cm.

87
LAMY, BERNARD (1640–1715)
TRAITEZ DE MECHANIQUE, DE L'EQUILIBRE, DES SOLIDES, ET DES LIQUEURS. Nouvelle édition, où l'on ajoûte une nouvelle maniére de démontrer les principaux théoremes de cette science. A Paris: André Pralard, 1687.
[ii], 163 (i.e. 263) [1], 23 [4] pp.; illus.; 1 leaf of plates (folding); 15 x 8 cm.

88
VIVIANI, VINCENZO (1622–1703)
DISCORSO AL SERENISSIMO COSIMO III GRANDVCA DI TOSCANA INTORNO AL DIFENDERSI DA' RIEMPIMENTI, e

dalle corrosioni de' fiumi applicato ad Arno in vicinanza della città di Firenze. In Firenze: Stamperia di Piero Matini, 1688.
[iv], 72 pp.; 26 x 18 cm.

89
BOUILLET,______ INGÉNIEUR (*fl.* 1693–1696)
TRAITÉ DES MOYENS DE RENDRE LES RIVIERES NAVIGABLES. . . . A Paris: Estienne Michallet, 1693.
[viii], 102 [2] pp.; 12 leaves of plates (7 folding); 20 x 14 cm.

90
[BOUILLET, ______ INGÉNIEUR (*fl.* 1693–1696)]
TRAITÉ DES MOYENS DE RENDRE LES RIVIERES NAVIGABLES. . . . A Amsterdam: Pierre Mortier, 1696.
[viii], 102, [+2] pp.; 12 leaves of plates (all folding); 19 x 11 cm.

91
FONTANA, CARLO (1634–1714)
UTILISSIMO TRATTATO DELL'ACQUE CORRENTI DIVISO IN TRE LIBRI, nel quale si notificano le misure, ed esperienze di esse. . . . In Roma: Stamparia di Gio. Francesco Buagni, 1696.
[xvi], 80 [+4], 81–196 pp. [+11 pp. index]; illus.; 2 leaves of plates (folding); 36 x 23 cm.

Well illustrated work on descriptive hydraulics.

92
HOSTE, PAUL (1652–1700)
L'ART DES ARMÉES NAVALES, OU TRAITÉ DES EVOLUTIONS NAVALES, qui contient des regles utiles aux officiers . . . d'une armée navale; avec des examples tirez de ce qui s'est passé de plus considérable sur la mer depuis cinquante ans. A Lyon: Anisson, & Posuel, 1697.
[xii], 424 pp.; 134 leaves of plates; 38 x 25 cm. Bound with the author's Théorie de la construction des vaisseaux (item 92 a).

92a
HOSTE, PAUL (1652–1700)
THÉORIE DE LA CONSTRUCTION DES VAISSEAUX, qui contient plusieurs traitez de mathématique sur des matiéres nouvelles & curieuses. A Lyon: Anisson, & Posuel, 1697.
[xii] 172 [4] pp.; 11 leaves of plates (all folding); 38 x 25 cm. Bound with the author's L'art des armées navales . . . (item 92).

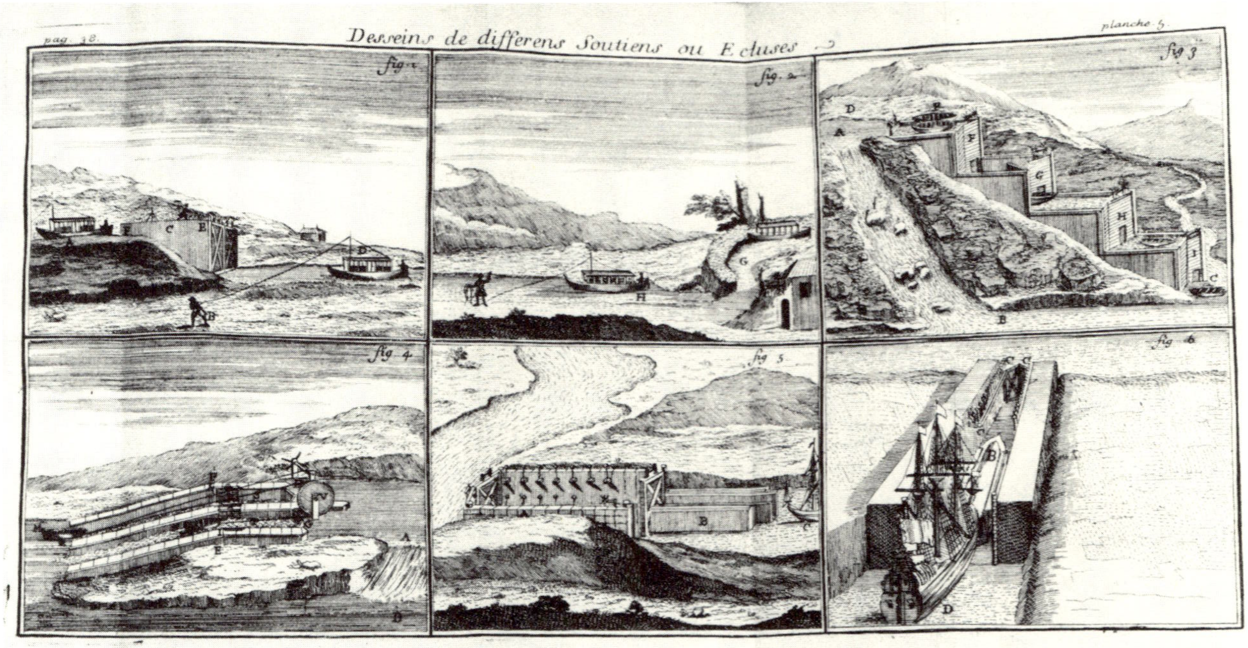

89. Types of river control discussed by Bouillet.

91. Fontana's study of liquid jets.

93
GUGLIELMINI, DOMENICO (1655–1710)
DELLA NATURA DE'FIUMI, TRATTATO FISICO-MATEMATICO. Bologna: Eredi d'Antonio Pisarri, 1697.
[xii], 375 pp. [1 p. errata]; 15 leaves of plates (all folding); 22 x 17 cm.

Professor of mathematics and then of hydrometry at the University of Bologna, as an authority on open-channel flow Guglielmini is considered by many to have founded the Italian school of hydraulics. This was the first noteworthy treatise on river hydraulics.

94
GUGLIELMINI, DOMENICO (1655–1710)
DELLA NATURA DE'FIUMI. . . . Nuova edizione. Annotazioni di Eustachio Manfredi. In Bologna: Stamperia di Lelio dalla Volpe, 1739.
[viii] xvi, 427 [1] pp.; 18 leaves of plates (all folding); 30 x 22 cm.

95
GUGLIELMINI, DOMENICO (1655–1710)
DELLA NATURA DE'FIUMI. . . . Nuova edizione con aggiunte. Annotazioni di Eustachio Manfredi. Milano: Società Tipog. de'Classici Italiani, 1852.
Vol. 1: [iii] iv–xxix, 271 pp. [+1 p. index]; frontis.; 5 leaves of plates (all folding); 21 x 13 cm.
Vol. 2: [iii] 4–318 pp.; 6 leaves of plates (all folding); 21 x 13 cm.

96
GUGLIELMINI, DOMENICO (1655–1710)
. . . OPERA OMNIA. . . . Accessit vita autoris a Jo. Baptista Morgagni. Genevae [Geneva]: Sumptibus Cramer, Perachon & Socii., 1719.
Vol. 1: [ii] 9 [+3], 772 pp.; 29 leaves of plates (18 folding); 23 x 17 cm.
Vol. 2: [ii], 571 pp.; frontis.; 23 x 15 cm.

Though he was a fine hydraulic engineer, Guglielmini's later writings were in the field of medicine.

97
FALCK, JOSEPH (*d.* 1737), PRAESES
LIQUIDORUM GRAVITAS ET AEQUILIBRIUM. Friburgi Brisgoiae [Freiburg im Breisgau]: Typis Joannis Jacobi Handler Typograph., 1713.
[xvi], 199 pp. [+1 p. errata]; 3 leaves of plates; 14 x 8 cm. A disserta-

tion presented at the University of Friburg by Ferdinando Fridolino Bürgin.

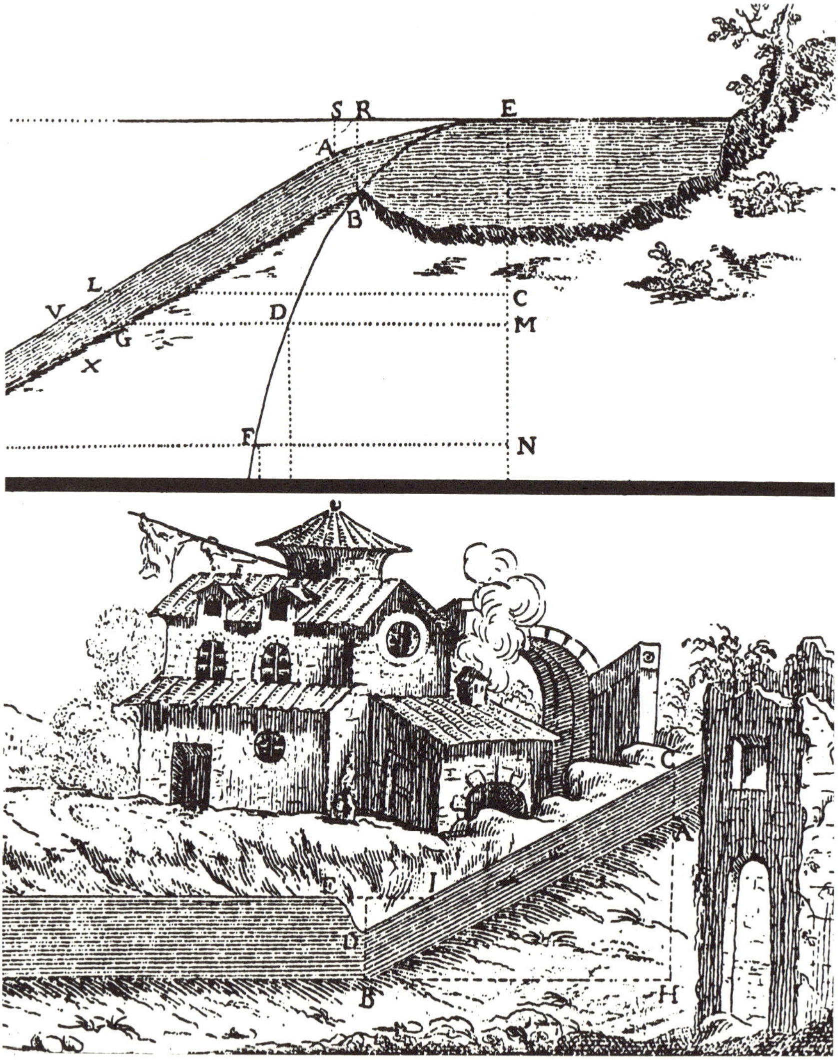

94. Typical sketches by Guglielmini of free-surface configuration.

98
Ditton, Humphrey (1675–1715)

THE NEW LAW OF FLUIDS: or, a discourse concerning the ascent of liquors, in exact geometrical figures, between two nearly contiguous surfaces to which is added the true state of the case about matter's thinking. . . . London: Printed by J. Roberts, for Benj. Cowse, 1714.

[viii] vii pp. [+1 p. adv.], 68 pp.; 73 [+6] pp. [+2 pp. adv.]; illus.; 19 x 11 cm.

99
BERNOULLI, JEAN (1667–1748)
ESSAY D'UNE NOUVELLE THEORIE DE LA MANOEUVRE DES VAISSEAUX, avec quelques lettres sur le même sujet. A Basle: Jean George König, 1714.

[xvi], 220 [4] pp.; 10 leaves of plates (all folding); 19 x 11 cm.

100
BERNOULLI, JEAN (1667–1748)
. . . OPERA OMNIA, TAM ANTEA SPARSIM EDITA, QUAM HACTENUS INEDITA. . . . Lausannae & Genevae [Lausanne & Geneva]: Sumptibus Marci-Michaelis Bousquet & Sociorum, 1742.

Vol. 1: [ii], xxiv, 563 pp.; frontis.; 23 leaves of plates (all folding); 24 x 18 cm.

Vol. 2: [v] 4–620 pp.; 17 leaves of plates (all folding); 24 x 18 cm.

Vol. 3: [v] 4–563 pp.; 36 leaves of plates (all folding); 24 x 18 cm.

Vol. 4: [iii] 6–588 pp.; 15 leaves of plates (all folding); 24 x 18 cm.

Brother of Jakob and father of Daniel, Johann was for a time the world's leading mathematician. As the result of jealousy over his son's Hydrodynamica, he predated his own subsequent book Hydraulica (see story in preface of item 101).

101
BERNOULLI, DANIEL (1700–1782) and JOHANN BERNOULLI (1667–1748)
HYDRODYNAMICS, BY DANIEL BOURNOULLI & HYDRAULICS, BY JOHANN BERNOULLI. Translated from the Latin by Thomas Carmody and Helmut Kobus. Preface by Hunter Rouse. New York: Dover Publications, Inc., [1968].

[v] viii–xv [+ix], 456 pp.; illus.; frontis.; 23 x 15 cm.

102
HERMANN, JACOB (1678–1733)
PHORONOMIA, SIVE DE VIRIBUS ET MOTIBUS CORPORUM SOLIDORUM ET FLUIDORUM LIBRI DUO. Amstelaedami [Amsterdam]: apud Rod. & Gerh. Wetstenios H.FF., 1716.

[xx], 401 pp. [2 pp. errata]; frontis.; 12 leaves of plates (all folding); 24 x 19 cm.

103
[WAHL, FERDINAND FRANÇOIS, COMTE DE]
TRAITTÉ DE L'ELEVATION DES EAUX. Enseignant ses principes & des maximes pour connoître les justes proportions qui sont à observer dans la conduite des eaux. . . . Munic: Mathieu Riedl, 1716.
[xxxii], 232 [+12] pp.; frontis.; 11 leaves of plates (all folding) and 27 leaves of tables (17 folding and 5 double); 25 x 19 cm. Includes: Essay des machines hydrauliques dans le quel est enseigne la maniere de mesurer les eaux courantes. . . .

104
POLENI, Giovanni, MARCHESE (1683–1761)
. . . DE MOTV AQVAE MIXTO LIBRI DVO. Quibus multa nova pertinentia ad aestuaria, ad portus, atque ad flumina continentur. Patavii [Padua]: Typis Iosephi Comini, 1717.
[viii], 132 [+3] pp.; 3 leaves of plates (all folding); 26 x 18 cm.

Poleni was professor of astronomy at Padua, then of physics, and finally of mathematics, as well as an engineer consultant. His name is usually associated with equations for flow through orifices and over weirs.

105
POLENI, GIOVANNI, MARCHESE (1683–1761)
DE CASTELLIS PER QVAE DERIVANTVR FLVVIORVM AQVAE HABENTIBVS LATERA CONVERGENTIA LIBER. Quo etiam continentur nova experimenta ad aquas fluentes, & ad percussionis vires pertinentia. Patavii [Padua]: Typis Iosephi Comini, 1718.
[vi], 65 [4] pp.; illus.; 1 leaf of plates; 27 x 18 cm.

106
POLENI, GIOVANNI, MARCHESE (1683–1761)
EPISTOLARVM MATHEMATICARVM FASCICVLVS. Patavii [Padua]: [Ex Typographi Seminaii], 1729.
[x], [186] [2] pp.; 11 leaves of plates (9 folding); 28 x 20 cm. Bound with the author's De mathesis in rebvs physicis vtilitate praelectio, 1720.

107
STURM, LEONHARD CHRISTOPH (1669–1719)
GRÜNDLICHE UND PRACTISCHE UNTERWEISUNG, wie man Fang-Schläussen und Roll-Brücken, nach der besten heutigen Art, von

103. Jets and fountains treated by Wahl.

105. Experimental hydraulics after Poleni.

Holtz und Stein, starck, beständig und bequem bauen solle. . . . Augspurg: In Verlegung Jeremiae Wolffens, 1720.

[28] pp.; 8 leaves of plates; 33 x 20 cm.

108

RACCOLTA D'AUTORI CHE TRATTANO DEL MOTO DELL'ACQVE. In Firenze: Nella stamperia di S.A.R. Per gli Tartini, e Franchi, 1723.

Vol. 1: [iv] v–xxxvii [+1], 408 pp.; illus.; 10 leaves of plates (all folding); 24 x 17 cm.

Vol. 2: [vi], 3–713 pp.; illus.; 25 leaves of plates (all folding); 24 x 17 cm.

Vol. 3: [vi], 3–576 pp.; illus.; 11 leaves of plates (all folding); 24 x 17 cm.

Volume I reviews works by Archimedes, Galileo, Castelli, Michelini, Viviana, Borelli, and others; II, works by Guglielmini and Grandi; III, Manfredi, Piccard and Narducci.

109

JURIN, JAMES (1684–1750)

. . . DISSERTATIONIS DE MOTU AQUARUM FLUENTIUM CONTRA NONNULLAS PETRI ANTONII MICHELOTTI ANIMADVERSIONES. . . . Venetiis [Venice]: Typis Pinellorum Fratrum, 1724.

[ii] 3–38 pp. [+1 p. errata]; 1 leaf of plates; 24 x 18 cm. Bound with Suzzius, Josephus. Disquisitiones mathematicae, and Ripa, Ludovicus A. Miscellanea.

110

JURIN, JAMES (1684–1750)

DE MENSURA ET MOTU AQUARUM FLUENTIUM. Tentamen primum. Quo agitur de aqua effluente ex vase semper pleno per foramen rotundem: et de resistentia ejusdem ex defectu fluiditatis oriunda. [London] [s.n.] [1739?].

[iv], 63 pp.; 1 leaf of plates (folding); 21 x 17 cm. Issued as Royal Society of London, Philosophical Transactions, Vol. 41, nos. 452 and 453 (1739).

111

LEUPOLD, JACOB (1674–1727)

THEATRI MACHINARUM HYDRAULICARUM . . . ODER: Schau-Platz der Wasser-Künste. Leipzig: Bey dem Autore und Joh.

Friedr. Gleditschens seel. Sohn, Vol. 1, 1724; Vol. 2, 1725.
Vol. 1: [xii], 172 pp. [+2 pp. index]; 53 leaves of plates; 39 x 24 cm.
Vol. 2: [xvii], 165 pp. [+3 pp. index]; 58 leaves of plates (2 folding); 39 x 24 cm.
Bound with other works by the author.

A tremendous compilation and illustration of contemporary hydraulic and related subjects.

112
VARIGNON, PIERRE (1654–1722)
TRAITÉ DU MOUVEMENT, ET DE LA MESURE DES EAUX COULANTES ET JAILLISSANTES. . . . Tiré des ouvrages manuscrits de feu Monsieur Varignon par M. l'abbé Pujol. A Paris: Pissot, 1725.
[xvi], 128 pp.; 5 leaves of plates (all folding); 25 x 18 cm.

Emphasis herein was placed on mechanics rather than hydraulics.

113
HOOKE, ROBERT (1635–1703)
PHILOSOPHICAL EXPERIMENTS AND OBSERVATIONS OF THE LATE EMINENT DR. ROBERT HOOKE . . . and other eminent virtuoso's in his time. [Edited by] W. Derham. London: Published by W. Derham. Printed by W. and J. Innys, 1726.
[viii], 391 pp. [+7 pp. index, +2 pp. adv.]; illus.; 4 leaves of plates (2 folding); 19 x 12 cm.

114
SWITZER, STEPHEN (1682?–1745)
AN INTRODUCTION TO A GENERAL SYSTEM OF HYDROSTATICKS AND HYDRAULICKS, PHILOSOPHICAL AND PRACTICAL. Wherein the most reasonable and advantageous methods of raising and conducting water, for the watering noblemens and gentlemens seats, buildings, gardens, &c. are carefully (and in a manner not yet publish'd in any language) laid down. . . . London: Printed for T. Astley. . . . and L. Gilliver, 1729.
Vol. 1: [vi], xxxii [+iv], 133 [+15.], [129] - 274, 10 pp.
Vol. 2: [viii], [275] -352, 4, [353]-413 [+14] pp.; frontis.; 59 leaves of plates (all folding); 25 x 18 cm. Two volumes in one.

115
NARDUCCI, TOMMASO, DA LUCCA (1679–1766)
LA QUANTITÀ DEL MOTO, O SIA LA FORZA DELL'ACQVE

CORRENTI, dimostrata ne'diversi stati, e supposti delle medesime. Lucca: Francesco Marescandoli, 1733.

[vi], 7–200 pp.; 4 leaves of plates (all folding); 23 x 16 cm.

116
CLARE, MARTIN (*d.*1751)

THE MOTION OF FLUIDS, NATURAL AND ARTIFICIAL; in particular that of the air and water, . . . London: Printed for Edward Symon, 1735.

[xx], 323 pp. [25 pp. glossary, index, errata, adv.]; 9 leaves of plates; 19 x 12 cm.

For its time, this was an excellent textbook.

117
CLARE, MARTIN (*d.*1751)

THE MOTION OF FLUIDS, NATURAL AND ARTIFICIAL; in particular that of the air and water: . . . Second edition, corrected and improved. London: Printed for Edward Symon, 1737.

[xvi], 369 pp. [23 pp. glossary, index, adv.]; 9 leaves of plates; 20 x 13 cm.

118
CLARE, MARTIN (*d.*1751)

THE MOTION OF FLUIDS, NATURAL AND ARTIFICIAL; in particular that of the air and water: . . . Third edition, corrected and improved. London: Printed for A. Ward, J. and P. Knapton, S. Birt, C. Hitch, J. Jodges, S. Austen, J. Osborn, J. Robinson, J. Davidson, J. Beecroft, R. Dodd, and J. and J. Rivington, 1747.

[xvi], 375 pp. [+25 pp. glossary, index, adv.]; 9 leaves of plates (all folding); 20 x 12 cm.

119
CLARE, MARTIN (*d.* 1751)

A TREATISE ON THE MOTION OF FLUIDS, NATURAL AND ARTIFICIAL; in which that of air and water is particularly considered, and demonstrated in a plain and familiar manner. Revised and corrected with considerable additions, by R. Hall. London: Printed by F. D. Dewick . . . for Vernor and Hood, 1802.

[iii] iv–viii [3], 380 pp.; 10 leaves of plates; 21 x 13 cm.

120
BELIDOR, BERNARD FOREST DE (1697?–1761)

ARCHITECTURE HYDRAULIQUE, OU L'ART DE CONDUIRE,

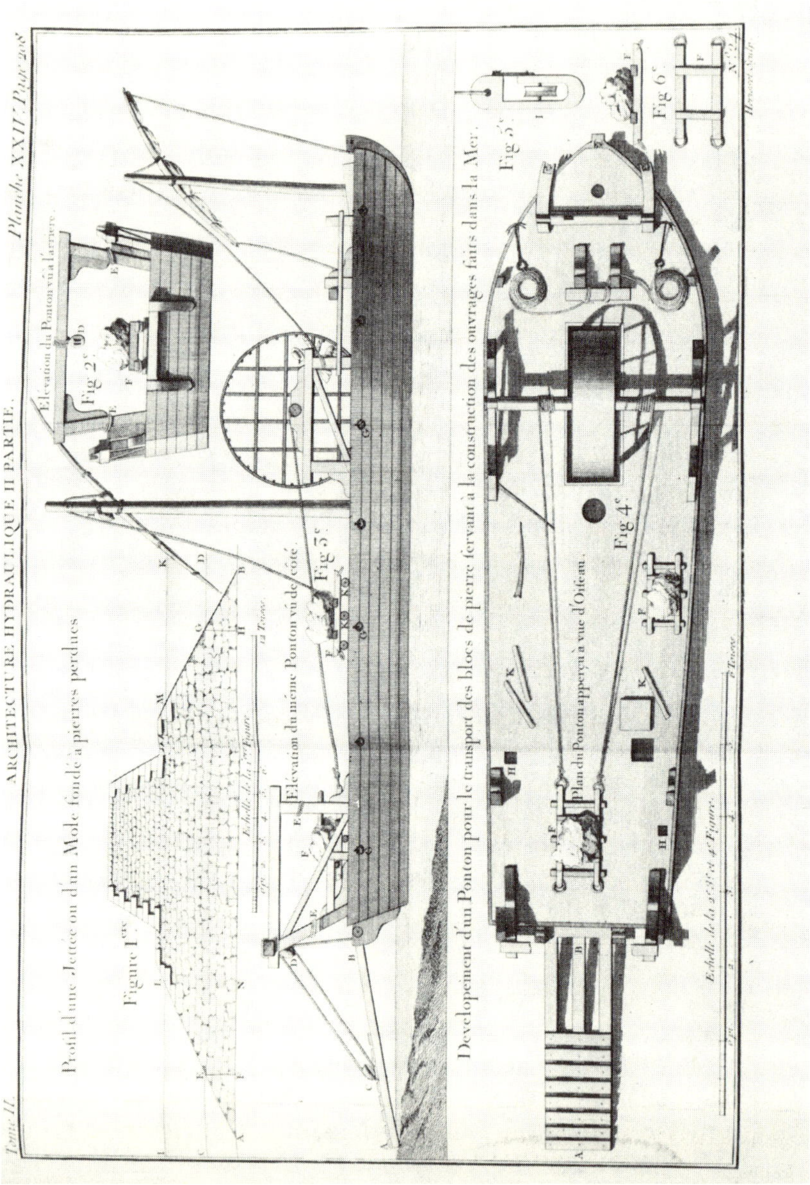

120. Dike building after Belidor.

D'ELEVER, et de menager les eaux pour les differens besoins de la vie. A Paris: Charles-Antoine Jombert, Vol. 1, 1737; Vol. 2, 1750; Vol. 3, 1739; Vol. 4, 1753.

Vol. 1: [viii], 412 pp.; frontis.; 43 leaves of plates (all folding); 28 x 21 cm.

Vol. 2: [viii], xiv, [+2], 412 [xxxii] pp.; 55 leaves of plates (all folding); 28 x 21 cm.

Vol. 3: [viii], xi, [v], 423 pp. [+1 p. errata], xxviii pp.; 60 leaves of plates (all folding); 28 x 21 cm.

Vol. 4: [iii] iv–viii, 480 pp. [xl pp.]; frontis.; 60 leaves of plates (all folding); 28 x 21 cm.

This was the French counterpart of the German Leupold (see item 111)—equally stupendous.

121

BELIDOR, BERNARD FOREST DE (1697?–1761)

ARCHITETTURA IDRAULICA OVVERO ARTE DI CONDURRE, innalzare e regolare le acque pei varj bisogni della vita. Con note ed aggiunte de Navier; versione Italiana su l'ultima edizione Francese di Basilio Soresina. Mantova: Fratelli Negretti, Vol. 1; 1835; Vol. 2, 1837; Vol. 3, 1838; Vol. 4, 1839.

Vol. 1: [v] vi–xxii, 276, 16, 320 pp.; 112 leaves of plates (55 double, 1 double folding); 28 x 21 cm.

Vol. 2: [v] vi–xii, 334 pp.; 109 leaves of plates (52 double, 2 double folding, 1 folding); 28 x 21 cm.

Vol. 3: [v] vi–xii, 354 pp.; 118 leaves of plates (52 double, 7 double folding); 28 x 21 cm.

Vol. 4: [iii] vi–viii, 367 pp.; 120 leaves of plates (59 double, 1 double folding); 28 x 21 cm.

(Scelta Biblioteca dell' Ingegnere Civile, vols. 10, 11, 12, [13].)

122

BERNOULLI, DANIEL (1700–1782)

HYDRODYNAMICA, SIVE DE VIRIBUS ET MOTIBUS FLUIDORUM COMMENTARII. Argentorati [Strasbourg]: Sumptibus Johannis Reinholdi Dulseckeri, 1738.

[viii], 304 pp.; 12 leaves of plates (all folding); 24 x 19 cm.

Probably the best-known work on hydraulics. Its author dealt perceptively with jet propulsion, manometry, rotating vessels, and even the kinetic theory of gases—but did not quite arrive at the Bernoulli theorem.

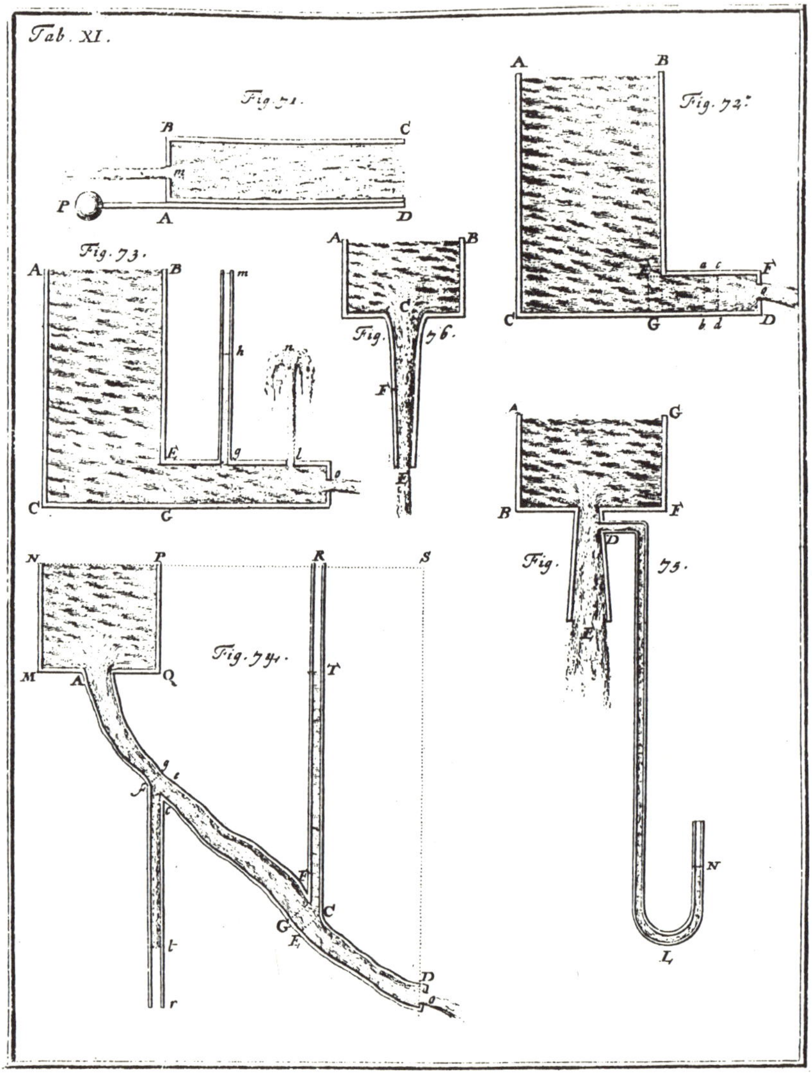

122. Typical Bernoullian efflux diagrams.

123
BERNOULLI, DANIEL (1700–1782) and JOHANN BERNOULLI (1667–1748)
HYDRODYNAMICS, BY DANIEL BERNOULLI & HYDRAULICS, BY JOHANN BERNOULLI. Translated from the Latin by Thomas Carmody and Helmut Kobus. Preface by Hunter Rouse. New York: Dover Publications, Inc., [1968].

[v] viii–xv [+ix], 456 pp.; illus.; frontis.; 23 x 15 cm.

DANIELIS BERNOULLI JOH. FIL.
MED. PROF. BASIL.
ACAD. SCIENT. IMPER. PETROPOLITANÆ, PRIUS MATHESEOS SUBLIMIORIS PROF. ORD. NUNC MEMBRI ET PROF. HONOR.

HYDRODYNAMICA,

SIVE

DE VIRIBUS ET MOTIBUS FLUIDORUM COMMENTARII.

OPUS ACADEMICUM

AB AUCTORE, DUM PETROPOLI AGERET, CONGESTUM.

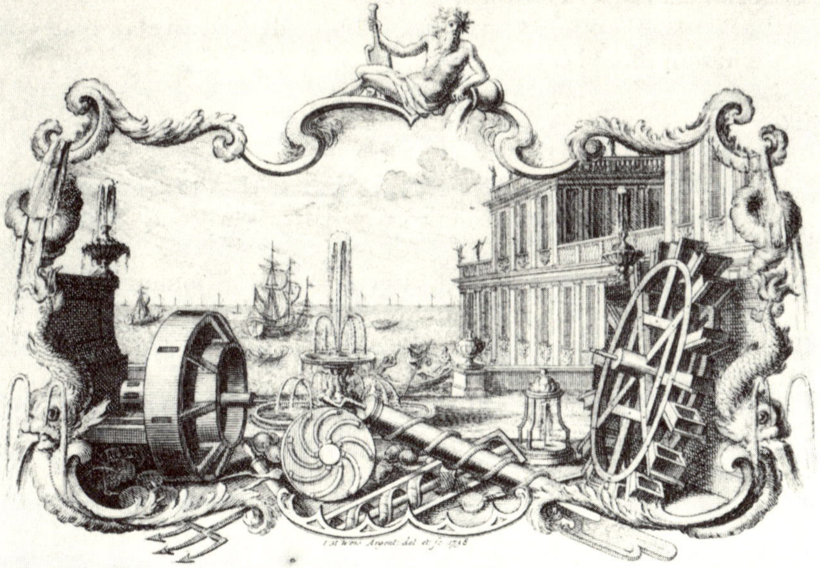

ARGENTORATI,

Sumptibus JOHANNIS REINHOLDI DULSECKERI,
Anno M D CC XXXVIII.

Typis JOH. HENR. DECKERI, Typographi Basiliensis.

122(2). Title page: Daniel Bernoulli's *Hydronamics*.

124
Cotes, Roger (1682–1716)
HYDROSTATICAL AND PNEUMATICAL LECTURES. Published with notes by his successor, Robert Smith. London: Printed for the editor, and sold by S. Austen, 1738.
[xvi], 243 pp. [+1 p. errata, +6 pp. index, +4 pp. adv.]; 5 leaves of plates (all folding); 20 x 12 cm.

Though written by the professor of astronomy and experimental philosophy at Cambridge, this book hardly ventured beyond hydrostatics.

125
Cotes, Roger (1682–1716)
HYDROSTATICAL AND PNEUMATICAL LECTURES. Second edition . . . by Robert Smith. Cambridge [England]: Printed by J. Bentham for W. Thurlbourn, 1747.
[xviii], 273 pp. [+8 pp. index, +2 pp. adv.]; 5 leaves of plates (all folding); 20 x 12 cm.

126
Cotes, Roger (1682–1716)
HYDROSTATICAL AND PNEUMATICAL LECTURES. Third edition, carefully revised and corrected. Notes by Robert Smith. London: Printed for J. Nourse, 1775.
[xvi], 288 pp. [+8 pp. index]; 5 leaves of plates (all folding); 21 x 13 cm.

127
Riva, Lodovico (1698–1746)
ANNOTAZIONI . . . AL TRATTATO DEL MOTO E DELLA MISURA delle acque correnti e dei getti delle medesime con un trattato preliminare del moto in generale, tratto dalle opere manuscritte del fu sig. Varignoni. . . . In Bologna: Lelio dalla Volpe, 1739.
[xii], 35 [+1], iii–vii [+i], 126 pp.; 4 leaves of plates (all folding); 25 x 17 cm. Contains Riva's translation into Italian of Varignon's Traité du mouvement, et de la mesure des eaux coulantes et jaillissantes . . . (see item 112).

128
Deidier, abbé (1696–1746)
LE MECHANIQUE GENERALE, CONTENANT LA STATIQUE, L'AIROMETRIE, L'HYDROSTATIQUE, ET L'HYDRAULIQUE,

pour servir d'indroduction aux sciences physico-mathematiques. A Paris: Charles-Antoine Jombert, 1741.

[viii], xxxiv [+2 pp. adv.], 632 pp.; 29 leaves of plates (all folding); 25 x 18 cm.

129
ZENDRINI, BERNARDINO (1679–1747)

LEGGI E FENOMENI, REGOLAZIONI ED USI DELLE ACQUE CORRENTI. Venezia: Presso Giambatista Pasquali, 1741.

[iv] v–xxxii, lix, 477 [+2] pp.; 12 leaves of plates (all folding); 27 x 19 cm.

130
ROBINS, BENJAMIN (1707–1751)

NEW PRINCIPLES OF GUNNERY: containing, the determination of the force of gun-powder, and an investigation of the difference in the resisting power of the air to swift and slow motions. London: Printed for J. Nourse, 1742.

[ii] iii–lvii, 95 pp.; 1 leaf of plates (folding); 20 x 12 cm.

131
ROBINS, BENJAMIN (1707–1751)

NEUE GRUNDSÄTZE DER ARTILLERIE. . . . Übersetzt und mit den nöthigen Erläuterungen und vielen Anmerkungen versehen, von Leonhard Euler. Berlin: A. Haude, 1745.

[xvi] 720 pp.; 8 leaves of plates (all folding); 18 x 10 cm.

Item 138 contains an English translation of Euler's elucidations and remarks upon Robins' work contained in this volume.

132
CLAIRAUT, ALEXIS CLAUDE (1713–1765)

THEORIE DE LA FIGURE DE LA TERRE, TIRÉE DES PRINCIPES DE L'HYDOSTATIQUE. A Paris: David, fils, 1743.

[iv] v–xl, 305 [+5 pp]; illus.; 13 x 20 cm.

133
D'ALEMBERT, JEAN LEROND (1717–1783)

TRAITÉ DE L'ÉQUILIBRE ET DU MOUVEMENT DES FLUIDES. Pour servir de suite au Traité de Dynamique. A Paris: David, l'aîné, 1744.

[ii] iii–xxxii, [viii], 458 pp. [+2 pp. errata]; 10 leaves of plates (all folding); 23 x 17 cm.

TRAITÉ
DE L'ÉQUILIBRE
ET DU MOUVEMENT
DES FLUIDES.

Pour ſervir de ſuite au Traité de Dynamique.

Par M. D'ALEMBERT, de l'Académie Royale des Sciences.

A PARIS,

Chez DAVID, l'aîné, Libraire, rue Saint Jacques, à la Plume d'or.

MDCCXLIV.

AVEC APPROBATION ET PRIVILEGE DU ROI.

133. A treatise by d'Alembert published in 1744.

Contemporary with the Bernoullis and Euler, the encyclopedist d'Alembert wrote on similar mathematical aspects of fluid motion.

134
D'ALEMBERT, JEAN LEROND (1717–1783)
TRAITÉ DE L'ÉQUILIBRE ET DU MOUVEMENT DES FLUIDES, pour servir de suite au Traité de dynamique. Nouvelle edition, revûe & augmentée par l'auteur. A Paris: Briasson, 1770.
[iv] v–xlvi [+1], 476 pp.; 10 leaves of plates (all folding); 22 x 16 cm.

135
D'ALEMBERT, JEAN LEROND (1717–1783)
ESSAI D'UNE NOUVELLE THEORIE DE LA RÉSISTANCE DES FLUIDES A Paris: David l'aîné, 1752.
[vi] vii–xlvi, 212 pp.; 2 leaves of plates (all folding); 21 x 15 cm.

Herein was presented the so-called paradox of the zero resistance of an immersed body in steady motion.

136
D'ALEMBERT, JEAN LEROND (1717–1783), MARIE JEAN ANTOINE NICOLAS CARITAT, MARQUIS DE CONDORCET (1743–1794), and CHARLES BOSSUT (1730–1814)
NOUVELLES EXPÉRIENCES SUR LA RÉSISTANCE DES FLUIDES. M. l'Abbé Bossut, rapporteur. A Paris: Claude-Antoine Jombert, 1777.
[iv] 232 pp.; 5 leaves of plates (all folding); 20 x 12 cm.

First towing-tank experiments, preceded only by those of Benjamin Franklin, described in 1768.

137
EULER, LEONHARD (1707–1783)
SCIENTIA NAVALIS SEV TRACTATVS DE CONSTRVENDIS AC DIRIGENDIS NAVIBVS. . . . Petropoli [St. Petersburg]: Typis Academiae Scientiarvm, 1749.
Vol. 1: [iv] 444 pp.; 37 leaves of plates (all folding); 24 x 19 cm.
Vol. 2: [ii] 534 pp.; 28 leaves of plates (all folding); 24 x 19 cm.

Successor to his teacher Johann Bernoulli as the world's leading mathematician, Euler wrote prolifically on all aspects of fluid motion, including the first derivation of the Bernoulli theorem—this despite the loss of sight in one eye and then in the other.

138
EULER, LEONHARD (1707–1783)
THE TRUE PRINCIPLES OF GUNNERY INVESTIGATED AND EXPLAINED. Comprehending translations of Professor Euler's Observations upon the new Principles of Gunnery, published by the late Mr. Benjamin Robins, and that celebrated Author's Discourse upon the Track described by a Body in a resisting Medium, inserted in the Memoirs of the Royal Academy of Berlin, for the Year 1753. . . . [Translated, with additions and remarks, by Hugh Brown]. London: Printed for J. Nourse, 1777.
[iii] iv–vii [+2], 366 [+2], xlvi pp.; 5 leaves of plates (all folding); 26 x 21 cm.

139
EULER, LEONHARD (1707–1783)
RECHERCHES SUR LE MOUVEMENT DES RIVIÈRES. Berlin: Haude et Spener, 1767.
[101]–118 pp.; 1 leaf of plates; 22 x 17 cm. A photocopy extract from: Histoire de l'Académie royale des sciences et belles-lettres [Berlin], Année MDCCLX [Vol. XVI].

140
EULER, LEONHARD (1707–1783)
THÉORIE COMPLETE DE LA CONSTRUCTION ET DE LA MANOEUVRE DES VAISSEAUX, mise à la portée de ceux qui s'appliquent à la navigation. Nouvelle édition corrigée & augmentée. A Paris: Claude-Antoine Jombert, 1776.
[viii] 268 pp.; 6 leaves of plates (all folding); 20 x 12 cm.

141
EULER, LEONHARD (1707–1783)
A COMPLETE THEORY OF THE CONSTRUCTION AND PROPERTIES OF VESSELS, with practical conclusions for the management of ships, made easy to navigators. New edition. Translated by Henry Watson. London: Printed for J. Sewell, 1790.
[iii] 4–8, xi, 281 pp.; frontis.; 11 leaves of plates (all folding); 21 x 14 cm. A translation of Euler's Théorie complete de la construction et de la manoeuvre des vaisseaux. See item 140.

142
EULER, LEONHARD (1707–1783)
DIE GESETZE DES GLEICHGEWICHTS UND DER BEWEGUNG FLÜSSIGER KÖRPER. Übersezt, mit einigen Abänderungen und

Zusätzen von H.W. Brandes. Leipzig: Siegfried Lebnecht Crusius, 1806.
[iii] iv–xxxii, 638 pp. [2 pp. errata]; 9 leaves of plates (all folding); 21 x 12 cm.

143
TRUESDELL, CLIFFORD AMBROSE (*b.* 1919)
EDITOR'S INTRODUCTION TO VOL. II, 13, OF EULER'S WORKS: I. The first three sections of Eulers [sic] treatise on fluid mechanics[,] 1766[.] II. The theory of aerial sound[,] 1687–1788[.] III. Rational fluid mechanics[,] 1765–1788[.] [Zurich]: [Orell Fussli], [1956].
[iii] viii–cxviii pp.; 29 x 21 cm. Offprint from Leonhardi Euleri Commentationes Mechanicae ad Theoriam Machinarum Pertinentes (Euleri Opera Omnia series II vol. 13).

144
ACKERET, JAKOB (1898–1981)
VORREDE ZU BAND II, 15, VON EULERS WERKEN: EULERS ARBEITEN ÜBER TURBINEN UND PUMPEN. Zürich: Orell Füssli, 1957.
[i] viii–lx [+1] pp.; illus.; 3 leaves of plates; 29 x 21 cm. Offprint from Leonhardi Euleri Commentationes Mechanicae ad Theoriam Machinarum Pertinentes (Euleri Opera Omnia series II Vol. 15).

145
STURM, LEONHARD CHRISTOPH (1669–1719)
VOLLSTÄNDIGE ANWEISUNG WASSER-KÜNSTE, WASSERLEITUNGEN, BRUNNEN UND CISTERNEN WOHL ANZUGEBEN. . . . Augspurg: Jeremias Wolff, 1752.
[ii] 3–18 pp.; illus.; 8 leaves of plates; 33 x 20 cm.

146
XIMENES, LEONARDO (1716–1786)
DISSERTAZIONE MECCANICA DI DUE STRUMENTI CHE POSSON SERVIRE ALLA GIUSTA STIMA DEL VIAGGIO MARITIMO E DELLA VELOCITÀ DELLE ACQUE, E DE'VENTI. Firenze: Stamperia imperiale, 1752.
[ii] iii–xxiii, 124 pp.; 4 leaves of plates (all folding); 22 x 14 cm.

The abate Ximenes was perhaps the most prolific of a series of Italian writers of this period, including Frisi, Gennette, Grandi, Lecchi, and Manfredi. Like the Raccolta, the writings were rather repetitious.

147
XIMENES, LEONARDO (1716–1786)
NUOVE SPERIENZE IDRAULICHE FATTE NE'CANALI, E NE'-FIUMI PER VERIFICARE LE PRINCIPALI LEGGI E FENOMENI DELLE ACQUE CORRENTI. Siena: Luigi e Benedetto Bindi, 1780.
[ii] ii–xxviii, 326 pp.; 2 leaves of plates (both folding); 29 x 22 cm.

148
XIMENES, LEONARDO (1716–1786)
PIANO DI OPERAZIONI IDRAULICHE PER OTTENERE LA MASSIMA DEPRESSIONE DEL LAGO DI SESTO O SIA DI BIENTINA. [Lucca]: [Francesco Bonsignori, 1782].
[ii] x, 347 pp.; 5 leaves of plates (all folding); 27 x 20 cm.

149
XIMENES, LEONARDO (1716–1786)
RACCOLTA DELLE PERIZIE ED OPUSCOLI IDRAULICI DEL SIGNOR ABATE LEONARDO XIMENES . . . alla quale si aggiungono le perizie di altri professori che anno scritto sulle stresse materie. Firenze: Pietro Allegrini, Vol. 1, 1785; Vol. 2, 1796.
Vol. 1: [ii] iii–xx, 472 pp.; 9 leaves of plates (all folding); 28 x 21 cm.
Vol. 2: [ii] iii–vi, 454 pp.; frontis.; 9 leaves of plates (all folding); 28 x 21 cm.

Includes contributions by Manfredi, Bertaglia, Perelli, and others.

150
FROMOND, GIOVANNI CLAUDIO (1703–1765)
DELLA FLUIDITÀ DE'CORPI. Livorno: Antonio Santini e Compagni, 1754.
[iii] iv–xxviii, 199 pp. [+1 p. errata]; 1 leaf of plates (folding); 21 x 15 cm.

151
BRAHMS, ALBERT
ANFANGS-GRÜNDE DER DEICH- UND WASSER-BAUKUNST, oder Gründliche Anweisung, wie man tüchtige haltbare Dämme wider die Gewalt der grössesten See-Fluthen bauen, auch dieselbe jederzeit im unvergeringerten Zustande erhalten könne . . . und was mehr bey Anlegung der Schleussen, Schlengen, Holzungen, Canäle und andern Wasser-Werken zu wissen nöthig und nützlich. . . . Aurich [Germany]: Hermann Tapper, Vol. 1, 1754?; Vol. 2, 1757.
Vol. 1: [xiv] 218 pp.; 21 x 17 cm.

Vol. 2: [viii] 131 pp. [+1 p. errata]; 6 leaves of plates (all folding); 21 x 17 cm. Vol. 2 has special title page: Anfangs-Gründe der Deich- und Wasser-Baukunst Anderer Theil. Vom Deich-Bau an Strömen und Flüssen. . . . Two volumes in one.

152
[Hébrard, Pierre]
CAMINOLOGIE, OU TRAITÉ DES CHEMINÉES, contenant des observations sur les différentes causes qui font fumer les cheminées, avec des moyens pour corriger ce défaut. A Dijon: F. Desventes, 1756.
[iv] xliv, 187 [17] pp.; frontis.; 20 leaves of plates (all folding); 17 x 10 cm.

153
Silberschlag, Johann Esaias (1721–1791)
ABHANDLUNG VOM WASSERBAU AN STRÖMEN. Leipzig: Johann Wendler, 1756.
[10] 246 pp.; 13 leaves of plates (all folding); 17 x 10 cm.

154
Silberschlag, Johann Esaias (1721–1791)
THÉORIE DES FLEUVES, AVEC L'ART DE BATIR DANS LEURS EAUX ET DE PRÉVENIR LEURS RAVAGES. [Translated from the German by "M. d'Au..Officier réformé."] A Paris: Charles-Antoine Jombert, 1769.
[iv] iii–ix [+1], 130 [+2] pp.; 13 leaves of plates (all folding); 28 x 21 cm.

155
Silberschlag, Johann Esaias (1721–1791)
AUSFÜHRLICHERE ABHANDLUNG DER HYDROTECHNIK ODER DES WASSERBAUES. Leipzig: Caspar Fritsch, Vol. 1, 1772 & Vol. 2, 1773.
Vol. 1: [xx] 422 pp. [+4 pp. adv.]; 16 leaves of plates (all folding); 19 x 12 cm.
Vol. 2: [ii] iii–xviii, 436 [+34] pp.; 14 leaves of plates (all folding); 19 x 12 cm.

156
Silberschlag, Johann Esaias (1721–1791)
AUSFÜHRLICHERE ABHANDLUNG DER HYDROTECHNIK ODER DES WASSERBAUES. Zweyte mit Anmerkungen vermehrte

Auflage. [Second edition with annotations added]. Wien: Johann Thomas Eblen von Trattnern, Vol. 1, 1785; Vol. 2, 1786.
Vol. 1: [xvi] 365 pp.; 16 leaves of plates (all folding); 20 x 12 cm.
Vol. 2: [iii] iv–xxvi [iv], 429 [+26] pp.; 14 leaves of plates (all folding); 20 x 12 cm.

157
KÄSTNER, ABRAHAM GOTTHELF (1719–1800)
MATHEMATISCHE ANFANGSGRÜNDE. Göttingen: Wittwe Vandenhoek, Vol. 1, 1758; Vol. 2, 1759; Vol. 3, 1760; Vol. 4, 1761; Vol. 5, 1766; Vol. 6, 1769.
Vol. 1: [xxiv] 423 [+1] pp.; 12 leaves of plates (all folding); 17 x 10 cm.
Vol. 2: [xxiv] 534 [+2] pp.; 14 leaves of plates (all folding); 17 x 10 cm.
Vol. 3: [xxiv] 433 [+1] pp.; 44 leaves of plates (all folding); 17 x 10 cm.
Vol. 4: [xxxii] 623 [+32] pp.; 4 leaves of plates (all folding); 17 x 10 cm.
Vol. 5: [xxiv] 424 pp.; 4 leaves of plates (all folding); 17 x 10 cm.
Vol. 6: [xxxiv] 618 pp.; 5 leaves of plates (all folding); 17 x 10 cm.

A series of books on mechanics by a Göttingen professor, one of them (see item 158) being on hydraulics.

158
KÄSTNER, ABRAHAM GOTTHELF (1719–1800)
ANFANGSGRÜNDE DER HYDRODYNAMIK WELCHE VON DER BEWEGUNG DES WASSERS BESONDERS DIE PRAKTISCHEN LEHREN ENTHALTEN. Zweite vermerhte Auflage. Göttingen: Vandenhoek und Ruprecht, 1797.
[xxviii] 692 pp.; 5 leaves of plates (all folding); 17 x 10 cm. (Mathematische anfangsgründe 4. Th., 2 Abth.)

159
SMEATON, JOHN (1724–1792)
AN EXPERIMENTAL ENQUIRY CONCERNING THE NATURAL POWERS OF WATER AND WIND TO TURN MILLS, and other machines, depending on a circular motion. London: n.p., 1760.
[iii] 4–77 pp.; 3 leaves of plates (all folding); 22 x 16 cm.

Smeaton not only conducted the first model tests on hydraulic machinery, but acquired prestige and membership in the Royal Society by virtue of his engineering works.

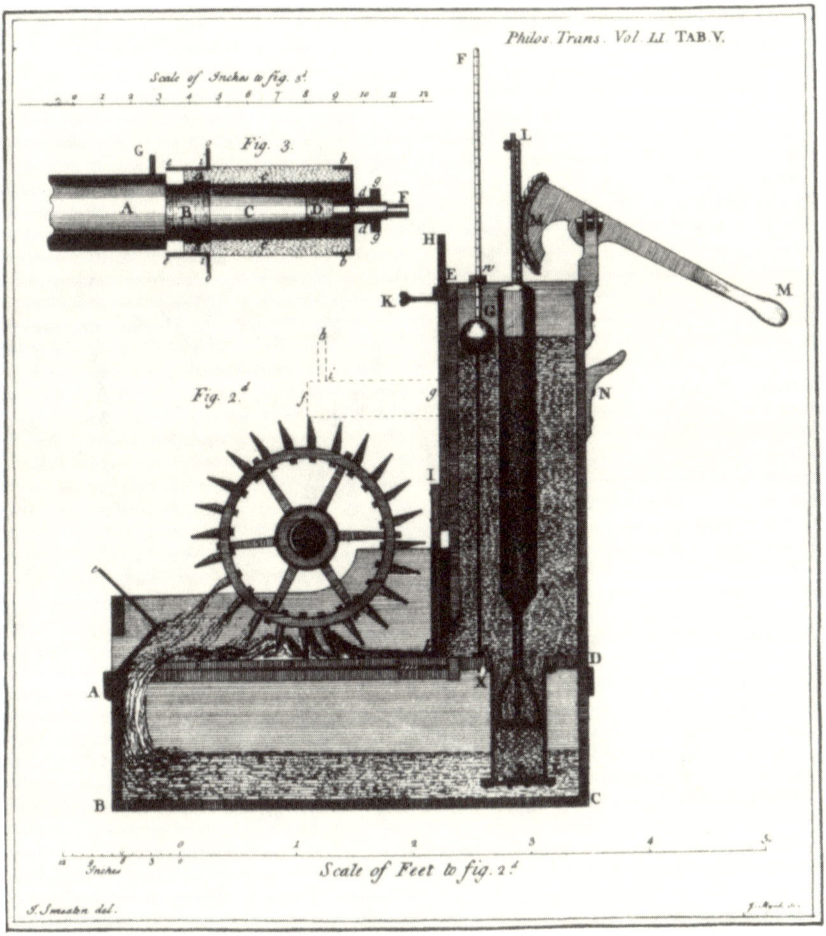

159. Smeaton's model of undershot wheel.

160
SMEATON, JOHN (1724–1792)

EXPERIMENTAL ENQUIRY CONCERNING THE NATURAL POWERS OF WIND AND WATER to turn mills and other machines depending on a circular motion. And an experimental examination of the quantity and proportion of mechanic power necessary to be employed in giving different degrees of velocity to heavy bodies from a state of rest. Also new fundamental experiments upon the collision of bodies. London: Printed for I. and J. Taylor, 1794.

[iii] iv, 110 pp.; 5 leaves of plates (all folding); 22 x 13 cm.

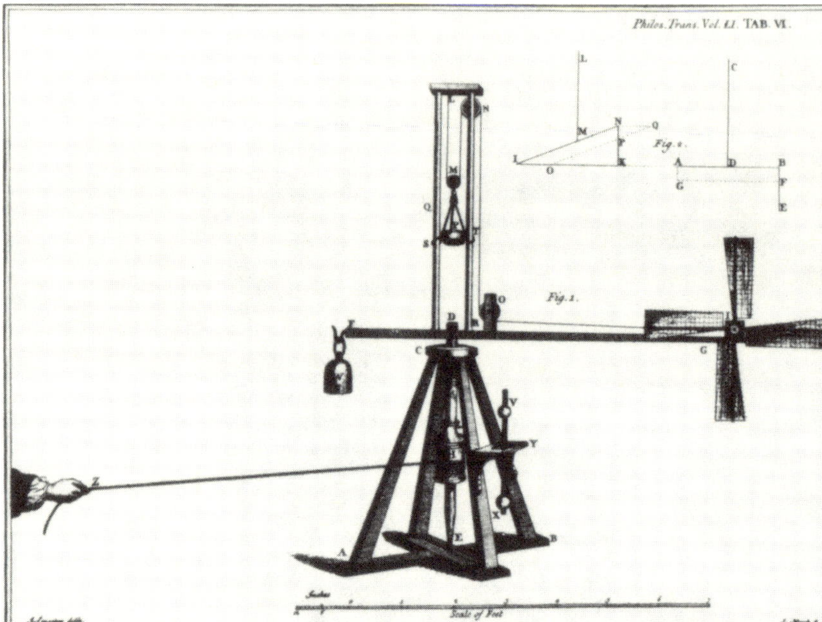

159(2). Smeaton's adaptation of the rotary arm for windmill tests.

161
SMEATON, JOHN (1724–1792)

EXPERIMENTAL ENQUIRY CONCERNING THE NATURAL POWERS OF WIND AND WATER to turn mills, and other machines, depending on a circular motion. And an experimental examination of the quantity and proportion of mechanic power necessary to be employed in giving different degrees of velocity to heavy bodies from a state of rest. Also new fundamental experiments upon the collision of bodies. Third edition. London: J. Taylor, 1813.

[iii] iv–vii, 110 pp. [+2 pp. adv.]; 5 leaves of plates (all folding); 25 x 15 cm.

162
SMEATON, JOHN (1724–1792)

RECHERCHES EXPÉRIMENTALES SUR L'EAU ET LE VENT, considérés comme forces motrices, applicables aux moulins et autres machines à mouvement circulaire, etc.; suivies d'expériences sur la transmission du mouvement et la collision des corps. [Translated from the English and with an introduction by P.S. Girard.] Paris: Courcier, 1810.

[iv] xxviii, 102 [+1] pp.; 5 leaves of plates (all folding); 26 x 20 cm.

163
Smeaton, John (1724–1792)
RECHERCHES EXPÉRIMENTALES SUR L'EAU ET LE VENT, considérés comme forces motrices applicables aux moulins et autres machines à mouvement circulaire, etc.; suivies d'expériences sur la transmission du mouvement et la collision des corps. Seconde édition. [Translated from the English and with an introduction by P.S. Girard.] Paris: Bachelier, 1827.

[v] vi–xxxii, 98 [+1] pp.; 5 leaves of plates (all folding); 27 x 21 cm.

164
Frisi, Paolo (1728–1784)
DEL MODO DI REGOLARE I FIUMI, e i torrenti, principalmente del Bolognese, e della Romagna. Lucca: Vincenzo Giuntini, 1762.

[ii] iii–viii, 172 pp. [1 p. errata]; 2 leaves of plates (both folding); 24 x 17 cm. Includes the author's Elogio del signore Gabriello Manfredi.

165
Frisi, Paolo (1728–1784)
DEI FIUMI E DEI TORRENTI. Aggiuntovi il trattato de'canali navigabili. Edizione terza accresciuta. Firenze: Stamperia Granducale, per Gaetano Cambiagi, 1770.

[ii] iii–iv, 140 pp. [+1 p. errata]; 2 leaves of plates (both folding); 28 x 21 cm.

166
Frisi, Paolo (1728–1784)
TRAITÉ DES RIVIÈRES ET DES TORRENS. Augmenté du Traité des Canaux navigables. [Translated from the Italian by Deserrey]. A Paris: Imprimerie Royale, 1774.

[vi] vii–xxiii [+i], 248 pp.; 3 leaves of plates (all folding); 26 x 20 cm.

167
Frisi, Paolo (1728–1784)
INSTITUZIONI DI MECCANICA, D'IDROSTATICA, D'IDROMETRIA E DELL 'ARCHITETTURA STATICA, E IDRAULICA ad uso della regia scuola eretta in Milano per gli architetti, e per gl'ingegneri. In Milano: Guiseppe Galeazzi Regio Stampatore, 1777.

[viii] 447 pp. [+1 p. errata]; 7 leaves of plates (all folding); 24 x 19 cm.

168
BORDA, JEAN CHARLES DE (1733–1799)
MÉMOIRE SUR L'ÉCOULEMENT DES FLUIDES PAR LES ORIFICES DES VASES.
579–607 pp.; illus.; 28 x 21 cm. In Mémoires de l'Académie royal des sciences. Paris: L'Académie, 1766. Photocopy.

169
MASCART, JEAN MARCEL (*b.* 1872.)
LA VIE ET LES TRAVAUX DU CHEVALIER JEAN-CHARLES DE BORDA. [Introduction by Emile Picard.] Lyon: A. Rey and Paris: A. Picard, 1919.
[viii] 821 pp.; 1 leaf of plates (folding); 24 x 15 cm. Annales de l'Universite de Lyon. Nouvelle Série. II. Droit, Lettres.—Fasicule 33.

170
VALLANCEY, CHARLES (1721–1812)
A TREATISE ON INLAND NAVIGATION, or, the art of making rivers navigable, of making canals in all sorts of soils, and of constructing locks and sluices. Extracted from the works of Guglielmini, Michelini, Castellus, Belidor, and others, with observations and remarks. Dublin: Printed for George and Alexander Ewing, 1763.
[v] iv–ix [i], 179 pp.; 24 leaves of plates (all folding); 27 x 19 cm.

171
BOSSUT, CHARLES (1730–1814) and VIALLET (engineer)
RECHERCHES SUR LA CONSUCTION LA PLUS AVANTAGEUSE DES DIGUES. A Paris: Charles-Antoine Jombert, 1764.
[v] 6–60 [4] pp.; 7 leaves of plates (all folding); 31 x 23 cm.

172
BOSSUT, CHARLES (1730–1814)
TRAITÉ ÉLÉMENTAIRE D'HYDRODYNAMIQUE: ouvrage dans lequel la théorie et l'expérience s'éclairent ou se suppléent mutuellement. . . . A Paris: Claude-Antoine Jombert, 1771.
Vol. 1: [ii] xxxvii [+i] 394 pp.; 9 leaves of plates (all folding); 20 x 12 cm.
Vol. 2: [ii], 444 pp. [+8 pp. adv.]; 7 leaves of plates (all folding); 20 x 12 cm.

The Abbé Bossut was the first to state expressly that engineering students should be taught the mechanics of fluids; his books—though good—were still hydraulics.

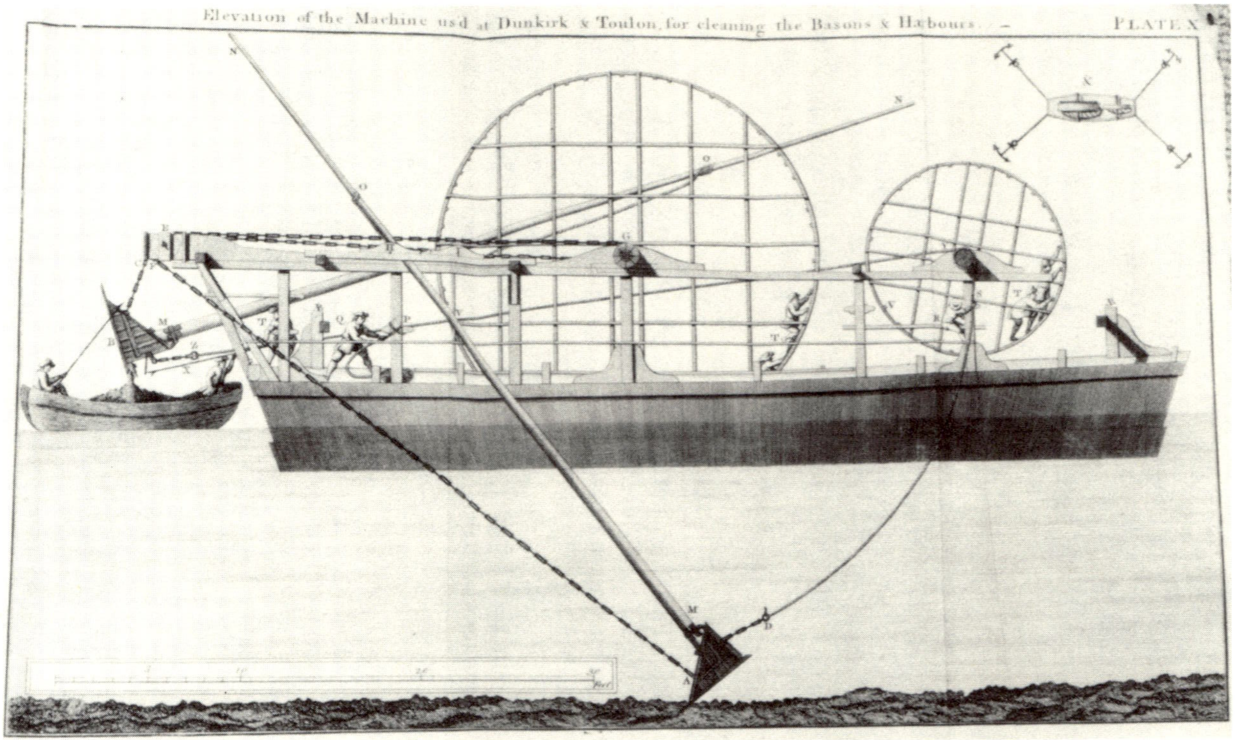

170. Details of dredge according to Vallancey.

173
BOSSUT, CHARLES (1730–1814)
TRAITÉ THÉORIQUE ET EXPÉRIMENTAL D'HYDRODYNAMIQUE. A Paris: Imprimerie Royale, 1786.
Vol. 1: [vi] xviii, 545 pp. [+3 pp. errata]; 12 leaves of plates (all folding); 22 x 13 cm.
Vol. 2: [viii] 515 pp. [+1 p. errata]; 11 leaves of plates (all folding); 22 x 13 cm.

174
BOSSUT, CHARLES (1730–1814)
TRAITÉ THÉORIQUE ET EXPÉRIMENTAL D'HYDRODYNAMIQUE. Nouvelle édition cor. et considérablement augmentée. A Paris: Laran, L'An IV [1796].
Vol. 1: [ii] 579 [5] pp.; 12 leaves of plates (all folding); 20 x 12 cm.
Vol. 2: [ii] 500 [4] pp.; 11 leaves of plates (all folding); 20 x 12 cm.

175
BOSSUT, CHARLES (1730–1814)
TRATTATO ELEMENTARE D'IDRODINAMICA. . . . Aggiuntevi Le lezioni d'idrodinamica del P. Gregorio Fontana. [Translated from the French by Giovanni Gratognini] In Pavia: R.I. Monistero di S. Salvatore, 1785.
[ii] iii–xii, 664 pp. [+4 pp. errata]; 6 leaves of plates (all folding); 18 x 12 cm. Also includes Teodoro Bonati's Saggio di une nuova teoria del movimento delle acque pei fiumi; and Appendice of P. Fontana.

176
REGI, FRANCESCO MARIA DE (1720–1794)
USO DELLA TAVOLA PARABOLICA NELLA MISURE DELLE ACQUE CORRENTI DESTINATE ALL'INNAFFIAMENTO DELLE TERRE. . . . In Milano: Giuseppe Richino Malatesta, Stampatore Regio Camerale, 1764.
[viii] 77 [+36] pp.; 1 leaf of plates (folding); 26 x 19 cm.

177
LECCHI, ANTONIO (1702–1776)
IDROSTATICA ESAMINATA NE'SUOI PRINCIPJ E STABILITA NELLE SUE REGOLE DELLA MISURA DELL'ACQUE CORRENTI. In Milano: Giuseppi Marelli, 1765.
[x] 459 pp.; illus.; frontis.; 27 x 20 cm.

IDROSTATICA
ESAMINATA
NE' SUOI PRINCIPJ
E STABILITA
NELLE SUE REGOLE
DELLA MISURA DELL' ACQUE
CORRENTI
DAL P. ANTONIO LECCHI
DELLA COMPAGNIA DI GESÙ
MATEMATICO DELLE LL. MM. II.

IN MILANO M.DCC.LXV.

NELLA STAMPERIA DI GIUSEPPE MARELLI
CON LICENZA DE' SUPERIORI.

177. Title page of Lecchi's *Hydrostatics*.

178
Lecchi, Antonio (1702–1776)
TRATTATO DE'CANALI NAVIGABILI. Seconda edizione. Milano: Giovanni Silvestri, 1824.
[v] vi–xvi, 365 pp. [+3 pp. adv.]; frontis.; 2 leaves of plates (folding); 15 x 9 cm. (Biblioteca scelta di opere italiane antiche e moderne. Classe VI.—Scienze ed arti. Vol. 140.)

179
NUOVA RACCOLTA D'AUTORI, CHE TRATTANO DEL MOTO DELL'ACQUE. In Parma: Filippo Carmignani, Vols. 1–6, 1766; Vol. 7, 1768.
Vol. 1: [x], xxii [+ii], 435 [+1] 13 leaves of plates (folding); 23 x 17 cm.
Vol. 2: [vi], 431, 129 pp.; 19 leaves of plates (all folding); 23 x 17 cm.
Vol. 3: [vi] 527 pp.; 13 leaves of plates (all folding); 23 x 17 cm.
Vol. 4: [vi] 447 pp.; 7 leaves of plates (all folding); 23 x 17 cm.
Vol. 5: [viii] 3–486 pp.; 11 leaves of plates (all folding); 23 x 17 cm.
Vol. 6: [vi] 4–391 pp.; 11 leaves of plates (all folding); 23 x 17 cm.
Vol. 7: [ix] 4–329, [xii] 110 pp.; 5 leaves of plates (all folding); 23 x 17 cm.

Volume 1, writings of Castelli, Galileo, Viviani, and Guglielmini; 2, Guglielmini and Manfredi; 3, Grandi and Poleni; 4, Castelli and Manfredi; 5, Zendrini; 6, Piccard, Gennete, Manfredi, Bolognini, Narducci, Grandi, and others; 7, Lecchi, Lorgna, Frisi, and others.

180
Michelotti, Francesco Domenico (1710–1777)
SPERIMENTI IDRAULICI PRINCIPALMENTE DIRETTI A CONFERMARE LA TEORICA, e facilitare la pratica del misurare le acque correnti. In Torino: Stamperia Reale, Vol. 1, 1767; Vol. 2, 1771.
Vol. 1: [xxiv], 206 pp. [+1 p. errata]; frontis.; 16 leaves of plates (all folding); 24 x 19 cm.
Vol. 2: [viii], 234 [+6] pp.; 2 leaves of plates (all folding); 24 x 19 cm. Title of Vol. 2 has slight variant: Sperimenti idraulici principalmente diretti a confermare la teorica, ed a facilitare la pratica. . . .

181
Michelotti, Francesco Domenico (1710–1777)
HYDRAULISCHE VERSUCHE ZUR BEGRÜNDUNG UND BEFÖRDEERUNG DER THEORIE UND PRACTIK. Nebst einem Anhange, welcher die neuesten Turiner Versuche von Joseph Terese

Michelotti enthält. [Translated by C.G. Zimmermann with annotations by J. A. Eytelwein.] Berlin: Im Verlage der Realschulbuchhandlung, 1808.

[v] vi–xxiii [+i], 253 pp. [+1 p. errata]; 4 leaves of plates (all folding); 24 x 20 cm.

182

LORGNA, ANTON MARIA (1730–1796)

DISCORSO INTORNO AL RIPARARE DALLE INONDAZIONI DELL'ADIGE LA CITTÀ DI VERONA. [Verona]; Stamperia [Marco] Moroni, 1768.

[vi] 7–55 pp.; 2 leaves of plates (both folding); 24 x 17 cm. Bound with other works by Lorgna, viz. items 182a, 182b.

182a

LORGNA, ANTON MARIA (1730–1796)

DEL MODO DI MIGLIORARE L'ARIA DI MANTOVA. In Verona: Marco Moroni, 1771.

[iv] 5–52 pp.; 24 x 17 cm. Bound with other works by Lorgna, viz. items 182, 182b.

182b

LORGNA, ANTON MARIA (1730–1796)

RICERCHE INTORNO ALLA DISTRIBUZIONE DELLE VELOCITÀ NELLE SEZIONI DE'FIUMI. In Verona: Marco Moroni, 1771.

[ii] 3–51 pp. [+1 p. errata]; 1 leaf of plates (folding); 24 x 17 cm. Bound with other works by Lorgna, viz. items 182, 182a.

183

LORGNA, ANTON MARIO (1730–1796)

MEMORIE INTORNO ALL'ACQUE CORRENTI. In Verona: Stamperia Moroni, 1777.

[iii] iv–xix, 91 pp. [1 p. errata]; 2 leaves of plates (both folding); 26 x 19 cm.

184

BINA, ANDREA (*b.* 1724)

RAGIONAMENTO SOPRA IL QUESITO QUAL SIA IL METODO PIÙ SICURO, più facile, e meno dispendioso tanto nell'esecuzione, che nella manutenzione, per impedire, e riparare la corrosione delle ripe de'fiume arginati, . . . In Mantova: Erede di Alberto Pazzoni, 1769.

[iv] 60 pp.; 3 leaves of plates (all folding); 29 x 20 cm. Bound with Anton Mario Lorgna's Dissertazione sopra il quesito essendo le pressioni dell'acqua stagnante in ragione delle altezze, . . . and Giovanni Verardo Zeviani's Dissertazione il sopra il quesito qual sia il preferibile fra i rimedi preservativi dallo scorbuto, . . .

185

VOCH, LUCAS (1728–1783)

EINLEITUNG ZU DER ARCHITECTURA HYDRAULICA: oder Gründlicher Unterricht, was man in dieser Wissenschaft von Brunnenfünsten . . . zu wissen nöthig hat. . . . Augsburg: n.p., 1769.

[v] 2–32 pp.; 6 leaves of plates (all folding); 31 x 18 cm.

186

BOURDET,

TRAITÉ DES DIGUES LE LONG DES FLEUVES ET DES RIVIERES, au quel on a joint des regles touchant les épis, fascinages et reversoirs &c. Berlin: Aux dépens de l'Auteur, 1771.

[ix] 10–164 [4] pp.; 9 leaves of plates (all folding); 16 x 9 cm.

187

JUAN Y SANTACILIA, JORGE (1713–1773)

EXAMEN MARITIMO THEÓRICO PRÁCTICO, Ó TRATADO DE MECHANICA APLICADO Á LA CONSTRUCCION, conocimiento y manejo de los navios y demas embarcaciones. Madrid: Imprenta de D. Francisco Manuel de Mena, 1771.

Vol. 1: [ii] iii–lviii, [xviii], 428 pp.; 5 leaves of plates (all folding); 20 x 14 cm.

Vol. 2: [xxxvi] 411 pp.; 8 leaves of plates (all folding); 20 x 14 cm.

First noteworthy hydraulics treatise in Spanish.

188

ROMPIASIO, GIULIO

METODO IN PRATICA DI SOMMARIO, O SIA COMPILAZIONE DELLE LEGGI, terminazioni, ed ordini appartenenti agl'illustrissimi ed eccellentissimi Collegio, e Magistrato alle Acque. . . . In Venezia: Stamperia Ducale, 1771.

[viii] xi–xxiv, 597 [5] pp.; 29 x 20 cm. Lacking port. and folding plates.

189
Deluc, Jean André (1727–1817)
RECHERCHES SUR LES MODIFICATIONS DE L'ATMOSPHÈRE. Contenant: l'histoire critique du baromètre et du thermometre, un traité sur la construction de ces instrumens, des experiences relatives à leur usages, et principalement à la mesure des hauteurs & à la correction des refractions moyennes. A Geneve: n.p., 1772.
Vol. 1: [v] ii–viii, 416 pp.; 7 leaves of plates (all folding); 26 x 20 cm.
Vol. 2: [iv] v–xii, 489 pp.; 26 x 20 cm.

190
Giannini, Petro
OPUSCULA MATHEMATICA. Parmae [Parma]: Ex Typographia Regia, 1773.
[iv] 214 pp.; 10 leaves of plates (all folding); 22 x 16 cm. Contents: I., De hydraulica. II., De cycloide contracta, ac protracta. III., De sectione determinata.

191
Karsten, Wencelaus Johann Gustav (1732–1787)
ABHANDLUNG ÜBER DIE VORTHEILHAFTESTE ANORDNUNG DER FEUERSPRÜTZEN . . . nebst noch einer Abhandlung über die allgemeine Theorie von der Bewegung des Wassers in Gefässen und Röhren von demselben Verfasser. Greisswald: A. F. Röse, 1773.
[iii] 4–215 [1] pp.; 5 leaves of plates (all folding); 23 x 18 cm.

192
Rocchi, Antonio (1724–1780)
DELLA MISURA DELLE VELOCITÀ, E DELLE FORZE DE'CORPI IN MOVIMENTO, APPLICATA ALLA MECCANICA, E ALLA IDROSTATICA. In Padova: Stamperia Conzatti, 1773.
[ii] 5–78 [+2] pp.; 1 leaf of plates (folding); 25 x 19 cm.

193
Belloni, Antonio (1736–1782)
DELL'ADIGE E DE'SUOI DIVERSIVI; TRATTATO FISICO-MATEMATICO. . . . In Venezia: Alvise Milocco, 1774.
[ii] iii–xii, 132 pp.; 2 leaves of plates (folding); 25 x 19 cm.

194
HERBERT, JOSEPH, EDLER VON, (1725–1794)
DISSERTATIO DE AQVAE, ALIORVMQVE NONNVLLORVM FLVIDORVM ELASTICITATE. Viennae: Typ. Jo. Thom. nob. de Trattnern, 1774.
[iii] 4–45 pp.; 1 leaf of plates (folding); 21 x 13 cm.

195
CAMETTI, OTTAVIANO (1711–1789)
MECHANICA FLUIDORUM, SIVE DE AEQUILIBRIO ET MOTU CORPORUM FLUIDORUM. Florentiae [Florence]: Ex typographia Stecchi et Pagani, 1777.
[ii] v–xvi, 237 pp.; 18 leaves of plates (all folding); 25 x 18 cm.

196
DU BUAT, PIERRE LOUIS GEORGES, COMTE (1734–1809)
PRINCIPES D'HYDRAULIQUE. . . . A Paris: Imprimerie de Monsieur, 1779.
[v] viii–xlviii, 397 [3] pp.; 7 leaves of plates (all folding); 20 x 13 cm.

The successive editions of Du Buat's work are noted not only for their excellence but also for the originality of his experiments on the resistance of conduits and immersed bodies—this despite his troubles during the Revolution.

197
DU BUAT, PIERRE LOUIS GEORGES, COMTE (1734–1809)
PRINCIPES D'HYDRAULIQUE. . . . Nouvelle édition. A Paris: Imprimerie de Monsieur, 1786.
Vol. 1: [v] vi–xlii, 453 pp.; 3 leaves of plates (all folding); 20 x 12 cm.
Vol. 2: [iv], 402 pp. [+5 pp. errata, adv.]; 1 leaf of plates (folding); 20 x 12 cm.

198
DU BUAT, PIERRE LOUIS GEORGES, COMTE (1734–1809)
PRINCIPES D'HYDRAULIQUE ET DE PYRODYNAMIQUE. . . . Nouvelle édition. A Paris: Firmin Didot, 1816.
Vol. 1: [v] vi–xxxvii, 448 pp.; 3 leaves of plates (all folding); 20 x 12 cm.
Vol. 2: [v] 6–352 pp.; 1 leaf of plates (folding); 20 x 12 cm.
Vol. 3: [v] vi–viii, 310 pp.; 8 leaves of plates (all folding) 20 x 12 cm.

199
Barré de Saint-Venant, Adhémar Jean Claude (1797–1886)
NOTICE SUR LA VIE ET LES OUVRAGES DE PIERRE LOUIS-GEORGES, COMTE DU BUAT. Lille: L. Danel, 1886.
[iv] 84 pp.; frontis.; 1 leaf of plates (folding); 23 x 14 cm.

200
Bettoni, Carlo
PENSIERI SUL GOVERNO DE'FIUMI. . . . Brescia: Pietro Vescovi, 1782.
[ii] 3–8, 312 pp. [+1 p. adv.]; 21 leaves of plates (all folding); 28 x 20 cm.

201
Barattieri, Giovanni Battista (1601–1677)
TRATTATO TEORICO-PRATICO CIRCA LA DIVISIONE DEGLI INCREMENTI FLUVIALI. . . . [Corrected, annotated and with a preliminary discourse by Francesco Sartorio.] Piacenza: Giuseppe Tedeschi, 1783.
[ii] iii–xvi, 203 [+1] pp.; 4 leaves of plates (all folding); 27 x 20 cm.

A discourse on the movement of river sediment. Reprint of Part 1, Bks. 3 & 4 of the author's Architettura d'acque.

202
Cocoli, Domenico (1747–1812)
DISSERTAZIONE SOPRA IL QUESITO STABILIRE LA VERA TEORIA DELLE ACQUE USCENTI DA'FORI APERTI NE'VASI, e mostrare in quai circostanze possa ella applicarsi alle acque correnti negli alvei naturali. . . . In Mantova: Alberto Pazzoni, 1783.
[ii] 3–120 pp. [+1 p. errata]; 1 leaf of plates (folding); 25 x 19 cm.

203
Fabre, Jean Antoine (1749–1834)
ESSAI SUR LA MANIERE LA PLUS AVANTAGEUSE DE CONSTRUIRE LES MACHINES HYDRAULIQUES ET EN PARTICULIER LES MOULINS A BLED; . . . A Paris: Alex. Jombert jeune, 1783.
[ix] ii–xvi, 402 pp.; 6 leaves of plates (all folding); 29 x 22 cm.

204
Fabre, Jean Antoine (1749–1834)
ESSAI SUR LA THÉORIE DES TORRENS ET DES RIVIÈRES,

CONTENANT LES MOYENS LES PLUS SIMPLES D'EN EMPÊCHER LES RAVAGES, d'en rétrécir le lit & d'y faciliter la navigation, le hallage & la flottaison. . . . A Paris: Bidault, An VI-1797.
[v] ii–xxxii, 284 pp.; 8 leaves of plates (all folding); 29 x 22 cm.

205
SAUSSURE, HORACE BÉNÉDICT DE (1740–1799)
ESSAIS SUR L'HYGROMÉTRIE. I Essai, Description d'un nouvel hygromètre comparable. II Essai, Théorie de l'hygrométrie. III Essai, Théorie de l'évaporation. IV Essai, Application des théories précédentes à quelques phénomenes de la météorologie. A Neuchatel: Samuel Fauche Pere et Fils, 1783.
[iv] xxiv, 367 pp.: 2 leaves of plates; 26 x 20 cm. Bound with the author's Défense de l'hygromètre à cheveu.

206
MARI, GIOSEFFO
L'IDRAULICA PRATICA RAGIONATA PROPOSTA A'SUOI DISCEPOLI. Guastalla: Regio-Ducale Stamperia di Salvatore Costa e Compagno, 1784.
Vol. 1: [xii] 171 pp.; 5 leaves of plates (all folding); 25 x 19 cm.
Vol. 2: [xiv] 166 pp.; 1 leaf of plates (folding); 25 x 19 cm.

207
MARI, GIOSEFFO
LE TEORIE IDRAULICHE CONCORDATE COLLE SPERIENZE PROPOSTE A' SUOI DISCEPOLI E AD USO DELLE SCUOLE. Guastalla: Regio-Ducale Stamperia di Salvatore Costa e Compagno, 1784.
Vol. 1: [xvi] 168 pp.; 1 leaf of plates (folding); 25 x 19 cm.
Vol. 2: [iv] 214 [+7] pp.; 1 leaf of plates; 25 x 19 cm. Two volumes in one.

208
LA CORDRAYE, FRANÇOIS CÉLESTIN DE LOYNES-BARRAUD, CHEVALIER DE, (ca. 1750–ca. 1815)
THÉORIES DES VENTS ET DES ONDES. . . . A Copenhague: Imprimerie de Niels Christensen, 1796.
[vii] 8–158 pp.; 1 leaf of plates (folding); 26 x ?cm. Photocopy.

209
CABRAL, ESTEVAM (1734–1811)
RICERCHE ISTORICHE, FISICHE ED IDROSTATICHE SOPRA

LA CADUTA DEL VELINO NELLA NERA COLLA DICHIARAZIONE DI UN NUOVO METODO PER DETERMINARE LA VELOCITÀ, e la quantità delle acque correnti ed altro nuovo metodo di elevare l'acque ne'sifoni a grande altezza. . . . In Roma: Antonio Fulgoni, 1786.

[ii] iii–xii, 82 pp.; 2 leaves of plates (both folding); 18 x 13 cm. Photocopy.

210
BERNARD, PONS-JOSEPH

NOUVEAUX PRINCIPES D'HYDRAULIQUE, APPLIQUÉS À TOUS LES OBJETS D'UTILITÉ, ET PARTICULIÈREMENT AUX RIVIÈRES, précédés d'un discours historique et critique sur les principaux ouvrages qui ont été publiés sur le même sujet. A Paris: Imprimerie de Didot l'Ainé, 1787.

[v] ii–lxvi, 331 pp.; 3 leaves of plates (all folding); 27 x 20 cm.

211
CANCRIN, FRANZ LUDWIG VON (1738–1812)

ABHANDLUNG VON DEM BAU DER WEHRE. Giesen [Giessen]: Krieger, 1788.

[iii] 4–176 pp.; 10 leaves of plates (all folding); 17 x 10 cm.

212
LAGRANGE, JOSEPH LOUIS, COMTE (1736–1813)

MÉCHANIQUE ANALITIQUE. A Paris: La Veuve Desaint, 1788.

[v] vi–xii, 512 pp.; 26 x 19 cm.; Section VII, Sur les principes de l'hydrodynamique; Section VIII, Du mouvement des fluides incompressibles; Section IX, Du mouvement des fluides compressibles et élastiques.

In this book, with no illustrations whatsoever, Lagrange—who had succeeded Euler as the world's greatest mathematician—further developed many mathematical concepts of flow originated by Euler and d'Alembert.

213
LAGRANGE, JOSEPH LOUIS, COMTE (1736–1813)

MÉCANIQUE ANALYTIQUE. [Annotated and corrected by M. J. Bertrand.] Troisième edition. Paris: Mallet-Bachelier, vol. 1, 1853; vol. 2, 1855.

Vol. 1: [v], ii–xiv; 428 pp.; 27 x 21 cm.

Vol. 2: [v], ii–iv, 390 pp. [+1 p. errata]; 27 x 21 cm. Two volumes in one.

214
MERCADIER, JEAN BAPTISTE (1750–1816)
RECHERCHES SUR LES ENSABLEMENS DES PORTS DE MER, et sur les moyens de les empêcher a l'avenir, particulierement dans les ports de Languedoc. A Montpellier: Imprimerie de Jean Martel ainé, 1788.
[iv] ii–xiv, 164 pp.; [2 pp. errata]; 10 leaves of plates (all folding); 24 x 18 cm.

215
SCHEMERL, JOSEPH VON (1752–1837)
ABHANDLUNG ÜBER DIE SCHIFFBARMACHUNG DER STRÖME. Wien: Johann Paul Kraus, 1788.
[xi] 12–196 pp.; 14 leaves of plates (all folding); 19 x 12 cm.

216
BRUNINGS, CHRISTIAAN
VERHANDELING OVER DE SNELHEID VAN STROOMEND WATER, en de middelen, om dezelve op allereïe diepten te bepaalen. Haarlem: J. van Walré, 1789.
[ii?] 3–233 pp.; 6 leaves of plates (2 folding); 21 x 13 cm. (Verhandelingen uitgegeeven door de Hollandsche Maatschappye der Weetenschappen, te Haarlem, deel 26.) Photocopy.

217
PESSUTI, GIOACCHINO (1743–1814)
OPUSCOLI DUE ALL'IDRODINAMICA APPARTENENTI IL PRIMO SULLA TEORIA DELLE TROMBE IDRAULICHE, il secondo sulla legge delle velocita' dell'acqua prorompente da'piccoli fori de' vasi. In Roma: Stamperia Pagliarini, 1789.
[iii] iv–vii [+iii], 188 pp.; 1 leaf of plates (folding); 24 x 18 cm.

218
PRONY, GASPARD CLAIR FRANÇOIS MARIE RICHE, BARON DE (1755–1839)
NOUVELLE ARCHITECTURE HYDRAULIQUE, contenant l'art d'élever l'eau au moyen de différentes machines, . . . et généralement de l'appliquer, de diverses manieres, aux besoins de la société. A Paris: Firmin Didot, Vol. 1, 1790; Vol. 2, 1796.

Vol. 1: [v] vi–xii, 621 pp. [3 pp. errata, adv.], 72 pp.; 15 leaves of plates (all folding); 31 x 23 cm.
Vol. 2: [v] 4–38, 203 pp. [1 p. errata]; 37 leaves of plates (all folding); 31 x 23 cm.

Finally director general of the École des ponts et chaussées, Prony taught, wrote, and practiced widely, especially on the subject of conduit resistance.

219
PRONY, GASPARD CLAIR FRANÇOIS MARIE RICHE, BARON DE (1755–1839)
MEMOIRE SUR LE JAUGEAGE DES EAUX COURANTES. A Paris: Imprimerie de la République, An X (1802 v. st.)
[ii] 68 pp.; 1 leaf of plates (folding); 25 x 20 cm.

220
PRONY, GASPARD CLAIR FRANÇOIS MARIE RICHE, BARON DE (1755–1839)
RECHERCHES PHYSICO-MATHÉMATIQUES SUR LA THÉORIE DES EAUX COURANTES. A Paris: Imprimerie Impériale, An XII (1804 v. st.)
[iii] ii–xxxii, 130 pp. [+1 p. errata]; 8 leaves of plates (all folding); 28 x 21 cm. Bound with the author's Recueil de cinq tables . . . (item 223).

221
PRONY, GASPARD CLAIR FRANÇOIS MARIE RICHE, BARON DE (1755–1839)
DESCRIPTION HYDROGRAPHIQUE ET HISTORIQUE DES MARAIS PONTINS. A Paris: Firmin Didot, Père et Fils, 1822.
[v] vi–xliv, 454 pp.: 25 x 20 cm. Companion atlas is item 222.

222
PRONY, GASPARD CLAIR FRANÇOIS MARIE RICHE, BARON DE (1755–1839)
ATLAS DES MARAIS PONTINS. Paris: Imprimerie de Firmin Didot, 1823.
[iii] iv–viii pp.; 78 leaves of plates (37 double, 2 double-folding); 43 x 27 cm. Companion text is item 221.

223
PRONY, GASPARD CLAIR FRANÇOIS MARIE RICHE, BARON DE (1755–1839)
RECUEIL DE CINQ TABLES, 1° pour faciliter et abréger les calculs des formules relatives au mouvement des eaux dans les canaux découverts et les tuyaux de conduite; 2° pour présenter les résultats de cent

soixante-sept expériences employées pour l'établissement de ces formules. Précédé d'une introduction et de l'exposition d'un procédé nouveau pour déterminer très-exactement, sans calculs ni opérations graphiques, et dans une grande longueur, une suite de points situés sur une même surface horizontale. A Paris: Imprimerie Royale, 1825.

[iii] 4–59 pp.; 1 leaf of plates (folding); 28 x 21 cm. Bound with the author's Recherches physico-mathématiques sur la théorie des eaux courantes (item 220).

224
PRONY, GASPARD CLAIR FRANÇOIS MARIE RICHE, BARON DE (1755–1839) and MOLARD

RAPPORT . . . SUR LES PROJETS, présentés au Comité des domaines et alienation, pour remplacer de machine de Marly. A Paris: Imprimerie Nationale, An III [1794]

[ii] 34 pp. [+1 p. errata]; 6 leaves of plates (3 colored, all folding); 27 x 17 cm.

225
PRONY, GASPARD CLAIR FRANÇOIS MARIE RICHE, BARON DE (1755–1839)

FORMULES POUR CALCULER LES HAUTEURS DES REMOUS. . . . Applications à des projets de grands travaux hydrauliques. A Paris: Carilian-Goeury, 1835.

[v] 6–24 pp.; 22 x 13 cm. Extracted from Annales des Ponts et Chaussées. 1st series, vol. 9 (1835) no. CCXXIV, pp. 237–256.

226
WOLTMAN, REINHARD (1757–1837)

THEORIE UND GEBRAUCH DES HYDROMETRISCHEN FLÜGELS oder Eine zuverlässige Methode die Geschwindigkeit der Winde und strömenden Gewässer zu beobachten. Hamburg: Benjamin Gottlob Hoffmann, 1790.

[ix] ii–iv [+iii] 60 pp. [+2 pp. errata]; 3 leaves of plates (all folding); 22 x 18 cm.

Best known for his prototype current meter, the harbor engineer Woltman also gave considerable attention to problems of fluid resistance.

227
WOLTMAN, REINHARD (1757–1837)

BEYTRÄGE ZUR HYDRAULISCHEN ARCHITECTUR. Göttin-

gen: Johann Christian Dieterich, Vol. 1, 1791; Vol. 2, 1792; Vol. 3, 1794; Vol. 4, 1799.
Vol. 1: [ix] x–xiv [+ii] 376 pp. [+3 pp. errata]; 2 leaves of plates (both folding); 20 x 12 cm.
Vol. 2: [iii] iv–xxxiv, [+ii] 304 pp. [+2 pp. errata]; 2 leaves of plates (both folding); 20 x 12 cm.
Vol. 3: [ix] x–lxii [+ii] 139 pp.; 4 leaves of plates (all folding); 20 x 12 cm.
Vol. 4: [ix] x–xvi [+iii] 4–424 pp.; 5 leaves of plates (all folding); 20 x 12 cm.

228
WOLTMAN, REINHARD (1757–1837)
BEYTRÄGE ZUR BAUKUNST SCHIFFBARER KANÄLE. Göttingen: Heinrich Dieterich, 1802.
[ix] x–xxiv, 382 pp.; 6 leaves of plates (all folding); 25 x 19 cm.

229
WOLTMAN, REINHARD (1757–1837)
BESCHREIBUNG DES HYDROMETRISCHEN FLÜGELS, und dessen Gebrauch als Wind- und Strom-Messer. Neue Auflage. Hamburg: August Campe, 1835.
[v] 6–48 pp.; 1 leaf of plates (folding); 23 x 24 cm.

230
LANGSDORF, KARL CHRISTIAN VON (1757–1834)
LEHRBUCH DER HYDRAULIK MIT BESTÄNDIGER RÜCKSICHT AUF DIE ERFAHRUNG. Altenburg: Richter, 1794.
[v] vi–cx, 655 pp.; 51 leaves of plates (all folding); 24 x 20 cm.

Comprehensive tome, profusely illustrated.

231
LANGSDORF, KARL CHRISTIAN VON (1757–1834)
UEBER DIE GESETZE DES STOSSES, vorzüglich in Anwendung auf den hydraulischen Stösser. . . . [Munchen] [G. Franz], [1811]
517–580 pp.; 6 leaves of plates (all double); 26 x 21 cm.

Early treatment of hydraulic ram.

232
CRISTIANI, GIROLAMO FRANCESCO
DELLA INALVEAZIONE, E DEL REGOLAMENTO DEL FIUME

BRENTA . . . trattato critico-idraulico. . . . Milano: Dalla Stamperia Camerale di Luigi Veladini, 1795.

[v] 6–365 pp. [1 p. errata]; frontis.; 1 leaf of plates (colored, folding); 26 x 20 cm.

233
Evans, Oliver (1755–1819)

THE YOUNG MILL-WRIGHT AND MILLER'S GUIDE. Philadelphia: Printed for and sold by the author, 1795.

[viii], 160, 178, 90, 10, [12] pp.; 26 leaves of plates. Photocopy (microcard).

Primitive application of hydraulics to mill construction.

234
Fulton, Robert (1765–1815)

A TREATISE ON THE IMPROVEMENT OF CANAL NAVIGATION; exhibiting the numerous advantages to be derived from small canals. . . . London: I. and J. Taylor, 1796.

[ii] iii–xvi, 144 pp.; 17 leaves of plates; 28 x 21 cm.

235
Baader, Joseph von (1763–1835)

VOLLSTÄNDIGE THEORIE DER SAUG- UND HEBE-PUMPEN, und Grundsätze zu ihrer vortheilhaftesten Anordnung, vorzüglich in Rücksicht auf Bergbau und Salinenwesen. . . . Bayreuth: Johann Andreas Lübeck, 1797.

[ix] x–lvi, 207 [+2] pp.; 6 leaves of plates (all folding); 25 x 20 cm.

236
Baader, Joseph von (1763–1835)

RAPPORT FAIT A LA PREMIERE CLASSE DE L'INSTITUT, le seize juin 1806, par Mm. Monge, Coulomb et Prony, sur le projet d'une nouvelle machine hydraulique, pour remplacer l'ancienne machine de Marly. A Paris: Imprimerie de Marguerite Chollet, 1807.

[v] 6–17 pp.; 26 x 21 cm.

237
Venturi, Giovanni Battiste (1746–1822)

RECHERCHES EXPÉRIMENTALES SUR LE PRINCIPE DE LA COMMUNICATION LATÉRALE DU MOUVEMENT DANS LES FLUIDES, appliqué à l'explication de differens phénomènes hydrauliques. A Paris, Houel et Ducros, An VI-(1797).

[iii] 4–88 pp.; 2 leaves of plates (both folding); 21 x 13 cm. Photocopy on microfilm.

Venturi, who divided his time between residence in Italy and in France, experimented not only with pipe section changes but also with pipe and open-channel resistance.

238
KENT, WALTER G.
AN APPRECIATION OF TWO GREAT WORKERS IN HYDRAULICS: Giovanni Battista Venturi, born 1746; Clemens Herschel, born 1842. London: Blades East & Blades, 1912.
[ii] 34 pp.; illus.; 25 x 19 cm.

239
WIEBEKING, KARL FRIEDRICH VON (1782–1842)
ALLGEMEINE AUF GESCHICHTE UND ERFAHRUNG GEGRÜNDETE THEORETISCH-PRACTISCHE WASSERBAUKUNST. Darmstadt: Johann Franz Peter Stahl, 1799.
[vii] ii–vii [+v], 502 pp.; 2 leaves of plates (both folding); 29 x 22 cm. Vol. 2 of 4 volumes.

Descriptive treatments of various aspects of hydraulic engineering.

240
EYTELWEIN, JOHANN ALBERT (1764–1848)
PRAKTISCHE ANWEISUNG ZUR KONSTRUKZION DER FASCHINENWERKE UND DEN DAZU GEHÖRIGEN ANLAGEN AN FLÜSSEN UND STRÖMEN nebst einer Anleitung zur Veranschlagung dieser Baue. Berlin: In Kommission bei Friedrich Maurer, 1800.
[iii] ix–x, 126 [+2] pp.; frontis.; 7 leaves of plates (5 folding); 26 x 21 cm.

The Prussian engineer Eytelwein paralleled the Frenchman Prony (see item 218) in both interests and practical accomplishments, including the formulation of resistance relationships.

241
EYTELWEIN, JOHANN ALBERT (1764–1848)
HANDBUCH DER MECHANIK FESTER KÖRPER UND DER HYDRAULIK. Mit vorzüglicher Rücksicht auf ihre Anwendung in der Architektur. Berlin: F.T. Lagarde, 1801.
[iii] iv–xxiv, 498 pp. [+1p. errata]; illus.; 5 leaves of plates (all folding); 20 x 12 cm.

242
Eytelwein, Johann Albert (1764–1848)
HANDBUCH DER MECHANIK FESTER KÖRPER UND DER HYDRAULIK. . . . Dritte, mit einem Anhange vermehrte Auflage. Leipzig: Friedrich Fleischer, 1842.
[iii] iv–xv, 414 pp. [+2 pp. errata]; illus.; 6 leaves of plates (all folding); 20 x 12 cm.

243
Eytelwein, Johann Albert (1764–1848)
BEMERKUNGEN ÜBER DIE WIRKUNG und vortheilhafte Anwendung des Stosshebers (Bélier hydraulique). . . . Berlin: [n.p.], 1805.
[iv] 102 pp.; 3 leaves of plates (all folding); 25 x 20 cm.

244
Eytelwein, Johann Albert (1764–1848)
HANDBUCH DER HYDROSTATIK, mit vorzüglicher Rücksicht auf ihre Anwendung in der Architektur. Berlin: G. Reimer, 1826.
[iii] iv–x, 202 pp.; 6 leaves of plates (all folding); 18 x 11 cm.

245
Crelle, August Leopold (1780–1855)
THEORIE DES WINDSTOSSES, welche in der Anwendung auf Windflügel, und die von denselben getriebenen Maschinen, durch eine völlige Uebereinstimmung mit der Erfahrung begrundet wird. Berlin: n.p., 1802.
[iii] 4–32 pp.; 1 leaf of plates; 24 x 20 cm.

246
Gilly, David (1748–1808) and Johann Albert Eytelwein (1764–1848)
PRAKTISCHE ANWEISUNG ZUR WASSERBAUKUNST, welche eine Anleitung zur Entwerfung, Veranschlagung und Ausführung der am gewöhnlichsten vorkommenden Wasserbaue enthält. Berlin: n.p., 2 vols. in 1, of 4. Vol. 1, 1802; Vol. 2, 1803.
Vol. 1: 14 leaves of plates (all folding); 25 x 18 cm.
Vol. 2: 14 leaves of plates (all folding); 25 x 18 cm.
Missing Vol. 3 (pub. 1805) and Vol. 4 (pub. 1808).

247
Girard, Pierre Simon (1765–1836)
ESSAI SUR LE MOUVEMENT DES EAUX COURANTES, et la

figure qu'il convient de donner aux canaux qui les contiennent. A Paris: Imprimerie de la République, An XII=1804.
[v] vi–vii, 60 pp.; 1 leaf of plates (folding); 29 x 21 cm.

Girard, engineer of the Ponts et Chaussées, contributed to the work of Prony, Woltman, and Eytelwein.

248
GIRARD, PIERRE SIMON (1765–1836)
NOTICE SUR LES JAUGES DE LA RIVIÈRE D'OURCQ ET DE SES AFFLUENS. A Paris: Imprimerie Impériale, Thermidor an XII [1804]
[ii] 16 pp.; 31 x 23 cm.

249
SCHEYER, JOHANN GEORG
PRAKTISCH-ÖKONOMISCHE WASSERBAUKUNST. [Vol. 1] Dritte verbesserte auflage. [Vol. 2] vermehrte und verbesserte zweite auflage. Leipzig: Vol. 1, Friedrich Fleischer, 1820; Vol. 2, Johann Benjamin Fleischer, 1803.
Vol. 1: [iii] iv–xii, 292 pp. [+1 p. adv.]; 19 leaves of plates (all folding); 20 x 12 cm.
Vol. 2: [iii] ii–xxiv, [iv] 304 pp.; 12 leaves of plates (all folding); 19 x 12 cm.

250
LECREULX, FRANÇOIS MICHEL (1734–1812)
RECHERCHES SUR LA FORMATION ET L'EXISTENCE DES RUISSEAUX, RIVIÈRES ET TORRENS qui circulent sur le globe terrestre; . . . A Paris: Bernard, An XII–1804.
[v] vi–xv, 407 pp. [+1 p. errata]; 8 leaves of plates (all folding); 25 x 19 cm.

251
MONTGOLFIER, JOSEPH MICHEL (1740–1810)
DE L'UTILITÉ DU BÉLIER HYDRAULIQUE. A Paris: L'Imprimerie de Gillé fils, An XIII (1805).
[ii] 18 pp.; 1 leaf of plates (folding); 18 x 12 cm.
Bound with two leaves entitled: "Notice-Pratique sur le Bélier hydraulique de M. Montgolfier."

252
Cessart, Louis Alexandre de (1719–1806)
DESCRIPTION DES TRAVAUX HYDRAULIQUES. Paris: Vol. 1, Collin, Renouard, Bernard, Magimel, 1806; Vol. 2, Collin, Renouard, Treuttel et Wurtz, Goeury, Magimel, 1808.
Vol. 1: [vii] ii–xiv, 316 pp.; frontis., 34 leaves of plates (33 folding); 29 x 21 cm.
Vol. 2: [v] ii–xvi, 358 pp.; 33 leaves of plates (all folding); 29 x 21 cm.

Extensively illustrated treatise by an inspector general of the Corps des Ponts et Chaussées.

253
Venturoli, Giuseppe (1768–1846)
ELEMENTI D'IDRAULICA. Bologna: Fratelli Masi e Compagno, 1807.
[iii] 2–4, 268 pp.; 2 leaves of plates (all folding); 20 x 12 cm.

Venturoli was professor of applied mathematics at the University of Bologna and eventually director of the engineering school at Rome. His textbook went through many editions. However, his greatest accomplishment was the derivation of the backwater equation in an anonymous paper, unfortunately not included in the collection.

254
Venturoli, Giuseppe (1768–1846)
RICERCHE SULLE RESISTENZE CHE RITARDANO LE ACQUE CORRENTI ed in particolare sulla resistenza d'attrito. In Modena: Presso La Societa' Tipografica, 1807.
[iv] 5–56 pp.; 1 leaf of plates (folding); 31 x 24 cm.

255
Venturoli, Giuseppe (1768–1846)
ELEMENTI DI MECCANICA E D'IDRAULICA. 3rd edition. Milano: Paolo Emilio Giusti, Vol. 1, 1817; Vol. 2, 1818.
Vol. 1: [xii] 416 pp.; 7 leaves of plates (all folding); 21 x 12 cm.
Vol. 2: [viii] 397 pp. [+2 pp. errata]; 21 x 12 cm.

256
Venturoli, Giuseppe (1768–1846)
ELEMENTI DI MECCANICA E D'IDRAULICA. 8th edition. Napoli: Saverio Starita, 1852. 2 vols. in 1.
Vol. 1: [iii] iv, 300 pp.; 7 leaves of plates (all folding); 21 x 12 cm.
Vol. 2: [iii] iv, 281 pp.; 7 leaves of plates (all folding); 21 x 12 cm.

257
BRÉMONTIER, NICOLAS THEODORE (1738–1809)
RECHERCHES SUR LE MOUVEMENT DES ONDES. A Paris: Firmin Didot, 1809.
[vii] vi–viii, 122 pp.; 6 leaves of plates (all folding); 21 x 14 cm.

Studies by an inspector general of the Corps des Ponts et Chaussées.

258
FISCHER, ERNST GOTTFRIED (1754–1831)
DARSTELLUNG UND KRITIK DER VERDUNSTUNGSLEHRE NACH DEN NEUESTEN BESONDERS DEN DALTONSCHEN VERSUCHEN. Berlin: Duncker und Humblot, 1810.
[iii] vi–viii, 86 pp. [+1 p. adv.]; 22 x 13 cm.

259
MENGOTTI, FRANCESCO, CONTE, (1749–1830)
SAGGIO SULL'ACQUE CORRENTI. Milano: Vol. 1, Tipi di Luigi Mussi, 1810; Vols. 2 & 3, Dalla Stamperia e fonderia di Gio. Gius. Destefanis, 1812. 3 vols. in 2.
Vol. 1: [iii] 4–142 [+1] pp.; 30 x 22 cm.
Vol. 2: [iii] 4–215 pp. [+1 p. errata]; 30 x 22 cm.
Vol. 3: [iii] 4–186 pp. [+1 p. errata]; 10 pp. of plates; 30 x 22 cm.

An extensive, largely qualitative, commentary on the flow of water, primarily in rivers.

260
MOLLET, JOSEPH (1756–1829)
HYDRAULIQUE PHYSIQUE, ou connaissance des phénomènes que présentent les fluides, soit dans l'état de repos, soit dans celui de mouvement, ouvrage élémentaire, renfermant l'hydrostatique et l'hydrodynamique. A Lyon: Ballanche père et fils, et Yvernault et Cabin, 1810.
[v] ii–xxiv, 482 pp. [+1 p. errata]; 11 leaves of plates (all folding); 20 x 12 cm. Presentation copy: "A l'illustre Baron de [?]ach, astronome de Gotha, De la part de l'auteur son très affectionné et obéissant serviteur."

261
ALDINI, GIOVANNI (1762–1834)
OSSERVAZIONI SUL FLUSSO DEL MARE, considerato come motore di molini e di altre macchine idrauliche. Milano: Stamperia Reale, 1811.
[iii] 4–28 pp.; 20 x 13 cm.

262
HACHETTE, JEAN NICOLAS PIERRE (1769–1834)
TRAITÉ ÉLÉMENTAIRE DES MACHINES. Paris: J. Klostermann fils, 1811.
[xi] viii–xix [+1 p. errata], 304 pp.; 29 leaves of plates (all folding); 26 x 19 cm.

Extensively illustrated treatise, largely on contemporary hydraulic machinery.

263
NEGRO, SALVATOR DAL (1768–1839)
ESPERIMENTI E CONSIDERAZIONI SULL'ARIETE IDRAULICO. Padova: Tipografia del Seminario, 1811.
[ii] 3–108 pp.; 2 leaves of plates; 29 x 21 cm.

264
PERIER, JACQUES CONSTANTIN (1742–1818)
MÉMOIRE SUR LA MACHINE DE MARLY. [Paris: Imprimerie de Porthmann, 1812?]
50 pp.; 27 x 21 cm.

Includes discussion of a steam engine as replacement.

265
BRANDES, HEINRICH WILHELM (1777–1834)
LEHRBUCH DER GESETZE DES GLEICHGEWICHTS und der Bewegung fester und flüssiger Körper. Leipzig: Paul Gotthelf Kummer, Vol. 1, 1817; Vol. 2, 1818.
Vol. 1: [iii] iv–xvi, 255 pp.; 5 leaves of plates (all folding); 21 x 12 cm. Photocopy.
Vol. 2: [iii] iv–xvi, 330 [+3 pp.]; 5 leaves of plates (all folding); 21 x 12 cm. Photocopy.

A typical German textbook by a professor of the University of Breslau.

266
PICCOLI, LUIGI
LE SERVITU PREDIALI, ridotte in casi pratici incisi in rame e geometricamente dimostrate. Nuova edizione. Milano: Batelli e fanfani, 1818.
[iii] iv–xx, [21]–59 pp.; 24 leaves of plates; 29 x 23 cm.

267
POISSON, SIMÉON DENIS (1781–1840)
MÉMOIRE SUR LA THÉORIE DES ONDES. . . . [Paris: Firmin Didot 1818.]
[71]–186 pp.; 27 x [?] cm. From Académie royale des sciences, Paris. Mémoires, 2. ser. t. 1. Photocopy.

One of the many interests of this mathematician.

268
COULOMB, CHARLES AUGUSTIN (1736–1806)
RECHERCHES SUR LES MOYENS D'EXÉCUTER SOUS L'EAU TOUTES SORTES DE TRAVAUX HYDRAULIQUES SANS EMPLOYER AUCUN ÉPUISEMENT. Troisième edition. A Paris: Bachelier, 1819.
[v] 6–40 pp.; 2 leaves of plates (both folding); 20 x 13 cm.

269
VINCE, SAMUEL (1749–1821)
THE PRINCIPLES OF HYDROSTATICS: designed for the use of students in the university. 5th edition. Cambridge [England]: Printed by J. Smith, Printer to the University, 1820.
[iv] 151 pp.; illus.; 21 x 13 cm.

270
BRUSCHETTI, GIUSEPPE
ISTORIA DEI PROGETTI E DELLE OPERE PER LA NAVIGAZIONE-INTERNA DEL MILANESE. Milano: Giovanni Bernardoni, 1821.
[iii] iv–xiv, 291 [+5] pp.; 6 leaves of plates (all folding); 25 x 19 cm.

An extensive discussion of canal systems in northern Italy.

271
RACCOLTA D'AUTORI ITALIANI CHE TRATTANO DEL MOTO DELL'ACQUE. Edizione quarta. Bologna: Jacopo Marsigli, Vol. 1, 1821; Vols. 2–5, 1822; Vols. 6–8, 1823; Vol. 9, 1824; Vol. 10, 1826.
Vol. 1: [x] iii–xiv, 366 pp.; 13 leaves of plates (all folding); 25 x 20 cm.
Vol. 2: [viii], 360 pp.; 4 leaves of plates (all folding); 25 x 20 cm.
Vol. 3: [iv] 5–452 pp.; 4 leaves of plates (all folding); 25 x 20 cm.
Vol. 4: [iv] 5–495 [1] pp.; 7 leaves of plates (all folding); 25 x 20 cm.
Vol. 5: [iv] v–lxii, [ii] 440 pp.; 5 leaves of plates (all folding); 25 x 20 cm.

Vol. 6: [iv], 5–539 pp.; 2 leaves of plates (all folding); 25 x 20 cm.

Vol. 7: [viii] 9–550 pp.; 2 leaves of plates (all folding); 25 x 20 cm.

Vol. 8: [iv] v–xxiii, 432 pp.; 12 leaves of plates (all folding); 25 x 20 cm.

Vol. 9: [iv] 5–559 pp.; 25 x 20 cm.

Vol. 10: [iv] v–vii, 464 pp.; frontis.; 53 leaves of plates (2 folding); 25 x 20 cm.

Volume 1, Guglielmini, "Della natura de'fiume"; 2, Guglielmini, "Della Misura dell'acque correnti": and letters; 3, Writings of Archimedes, Galileo, Castelli, Borelli, Torricelli, Viviani; 4, Grandi, Narducci, Michelini, Albizi, Montanaris; 5, Manfredi, and others; 6, Poleni, Frisi, Lecchi, and others; 7, Zanotti, Bosovich, Ximenes, Rappini; 8, Zendrini, "Leggi e regolazioni ed usi dell'acque correnti"; 9, Cassini and many others; 10, Manfredi, Zendrini, Poleni, Perelli, Ximenes, Chiesa and Gamberini, da Vinci.

272
FOURIER, JEAN BAPTISTE JOSEPH, BARON (1768–1830)
THE ANALYTICAL THEORY OF HEAT. Translated, with notes, by Alexander Freeman. New York: Dover Publications, Inc., [1955]
[v] vi–xxiii, 466 pp. [4 pp. adv.]; 20 x 13 cm.

First consideration of dimensional homogeneity.

273
NUOVA RACCOLTA D'AUTORI ITALIANI CHE TRATTANO DEL MOTO DELL'ACQUE. Bologna: Jacopo Marsigli, Vol. 1, 1823, Vols. 2–4, 1824; Cardinali e Frulli, Vol. 5, 1826; Francesco Cardinali, Vol. 6, 1829; Tipografia governativa—alle Volpe, Vol. 7, 1845.

Vol. 1: [viii] 9–304 pp.; 25 x 20 cm.

Vol. 2: [iv] 5–555, xxxii pp.; 12 leaves of plates (all folding); 25 x 20 cm.

Vol. 3: [viii] 365 pp.; 8 leaves of plates (all folding); 25 x 20 cm.

Vol. 4: [iv] 5–565 pp.; 3 leaves of plates (all folding); 25 x 20 cm.

Vol. 5: [ix] x–xv, 424 pp.; 30 leaves of plates (all folding); 25 x 20 cm.

Vol. 6: [viii] 9–490 [+1] pp.; 11 leaves of plates (all folding); 25 x 20 cm.

Vol. 7: [v] iv–xv, [iii] iv–vii, 273 [+3] pp.; 3 leaves of plates (all folding); 25 x 20 cm.

Volume I, Mengotti, "Idraulica fisica e sperimentale"; II, Writings of Lorgna, Zuliani, Focacci, Tadini, Bernareggi, Masetti; III, Fossombroni, Manetti; IV, Lecchi, Ferrari, Bruschetti, Parea; V, four-book miscellany; VI, various minor authors; VII, Corsini, Prony, Humbolt, Tadini, Fossombroni, Manetti, Libri, Guasti.

274

RICERCHE GEOMETRICHE ED IDROMETRICHE fatte nella scuola degl'ingegneri pontificj d'acque e strade, l'anno 1821. Milano: Paolo Emilio Giusti, 1822.

[iv] 59 pp.; 6 leaves of plates (all folding); 29 x 20 cm.

Reproduction of works by Venturoli, Eytelwein and others.

275

Tolotti, Antonio Sempiterni

ROTA IDROMETRICA, per misurare la velocita' delle acque correnti a' diverse profondita' ne' canali, torrenti, e fiumi, . . . Rieti: Tipografia di Luigi Bassoni, 1823.

[ii] 3–8 pp.; 1 leaf of plates; 27 x 19 cm.

276

Masetti, Giovanni Battista

DESCRIPTION, EXAMINATION, AND THEORY OF ALL HYDRAULIC TACHOMETERS PREVIOUSLY KNOWN. Translated by Dr. Carlo Zammattio and Arthur H. Frazier. [Unpublished typescript] [1968]

[6] 66 [+23] leaves; 35 leaves of plates; 28 x 21 cm. A translation of the descriptive portion of Masetti's article "Descrizione, esame e teoria di tutti i tachimetri idraulici" in Nuova Raccolta. . . . (item 273)

277

OPUSCOLI IDRAULICI RISGUARDANTI I CANALI NAVIGABILI NELL'ITALIA. . . . Bologna: Tipografia Marsigli, 1824.

[ii] 5–565 pp.; 3 leaves of plates (all folding); 26 x 20 cm.

A small version of the Raccolta with various works of Antonio Lecchi, Bernardino Ferrari, Giuseppe Bruschetti, Carlo Parea and G.B.M.

278

Pechmann, Heinrich Freiherr von (1773–1861)

PRAKTISCHE ANLEITUNG ZUM FLUSSBAUE. München: J. Lindauer, Vol. 1, 1825; Vol. 2, 1826.

Vol. 1: [iii] iv–xvi, viii, 199 pp. [+1 p. adv.]; 4 leaves of plates (all folding); 19 x 11 cm.

Vol. 2: [iii] iv–xvi, 278 pp. [+2 pp. adv.]; 4 leaves of plates (all folding); 19 x 11 cm.

A treatise on Bavarian river-hydraulic practice.

279
PONCELET, JEAN VICTOR (1788–1867)
MÉMOIRE SUR LES ROUES HYDRAULIQUES VERTICALES À AUBES COURBES, mues par-dessous, suivi d'expériences sur les effets mécaniques de ces roues. A Paris: Madame Huzard [et] Bachelier, 1825.
[ii] 56 pp.; 1 leaf of plates (folding); 27 x 21 cm. "Extrait du Bulletin de la Société d'Encouragement, Nos. CCLVII et CCLVIII. 24e année. Cahiers de Novembre et Décembre 1825."

280
PONCELET, JEAN VICTOR (1788–1867)
MÉMOIRE SUR LES ROUES HYDRAULIQUES À AUBES COURBES, mues par-dessous, suivi d'expériences sur les effets mécaniques de ces roues. . . . Nouvelle édition revue, corrigée et augmentée d'un second mémoire sur des expériences en grand relatives à la nouvelle roue, contenant une instruction pratique sur la manière de procéder à son établissement. Metz: Me Ve Thiel, 1827.
[v] vi–vii, 146 pp. [+2 pp. adv.]; 2 leaves of plates (both folding); 27 x 21 cm.

281
PONCELET, JEAN VICTOR (1788–1867) and JOSEPH AIMÉ LESBOS (1790–1860)
EXPÉRIENCES HYDRAULIQUES SUR LES LOIS DE L'ÉCOULEMENT DE L'EAU À TRAVERS LES ORIFICES RECTANGULAIRES VERTICAUX À GRANDES DIMENSIONS. A Paris: Imprimerie royale, 1832.
[iv] [i]–267 pp. [+1 p. errata]; 7 leaves of plates (all folding); 26 x 20 cm.

Oft-quoted measurements of flow through large, rectangular openings.

282
WEBER, ERNST HEINRICH (1795–1878) and WILHELM WEBER (1804–1891)
WELLENLEHRE AUF EXPERIMENTE GEGRÜNDET; oder, über

die Wellen tropfbarer Flüssigkeiten mit Anwendung auf die Schall- und Lichtwellen. Leipzig: Gerhard Fleischer, 1825.

[v] vi–xxviii, 574 pp. [+1 p. errata]; 18 leaves of plates (all folding); 20 x 13 cm.

Perceptive studies of wave motion by professors at Leipzig and Halle, respectively.

283
CAVALIERI SAN-BERTOLO, NICOLA (1788–1867)
ISTITUZIONI DI ARCHITETTURA STATICA E IDRAULICA. Bologna: Tipografia Cardinali e Frulli, Vol. 1, 1826; Vol. 2, 1827.

Vol. 1: [v] 6–384 pp.; 36 leaves of plates (all folding); 25 x 20 cm.

Vol. 2: [iii] 4–570 pp.; 32 leaves of plates (all folding); 25 x 20 cm.

Construction details for buildings, bridges, dredges, hydraulic machinery, etc., profusely illustrated, by a professor and engineer of the Corpo di Acque e Strade.

284
CAVALIERI SAN-BERTOLO, NICOLA (1788–1867)
ISTITUZIONI DI ARCHITETTURA STATICA E IDRAULICA. Napoli: Caro Batelli e Comp., 1840.

Vol. 1: [v] vi–vii, 177 pp. [+8 pp. index]; 36 leaves of plates (7 folding); 29 x 21 cm.

Vol. 2: [v] 6–286 pp.; 32 leaves of plates (20 folding); 29 x 21 cm.

285
CAVALIERI SAN-BERTOLO, NICOLA (1788–1867)
ISTITUZIONI DI ARCHITETTURA STATICA E IDRAULICA. [Enriched with notes and additions by Roberto Ferdinando Pisani.] Napoli: Raimondo Petraroja, Vol. 1, 1868; Vol. 2, 1869.

Vol. 1: [iii] vi–viii, 215 pp.; 29 x 20 cm.

Vol. 2: [v] 6–373 [1] pp.; 72 leaves of plates (31 folding); 29 x 20 cm. Two volumes in one.

286
ODDI, GIUSEPPI
SCHIARIMENTI ALLA MECCANICA ED IDRAULICA DEL SIG. PROFESSOR GIUSEPPE VENTUROLI. . . . Roma: Giuseppe Mauri, Vol. 1, 1826; Vol. 2, 1827.

Vol. 1: [ii] iii–vii [i] 395 pp. [+15 pp. errata]; 4 leaves of plates (all folding); 20 x 13 cm.

Vol. 2: [ii] 3–352 pp. [+17 pp. errata]; 2 leaves of plates (all folding); 20 x 13 cm.

287
GUDME, ANDREAS CASPAR

HANDBUCH DER THEORETISCHEN UND PRAKTISCHEN WASSERBAUKUNST. Berlin: August Rücker, Vol. 1, 1827; Vol. 2, pt. 1, 1828; Vol. 2, pt. 2, 1829; Vol. 3, 1829.

Vol. 1: [iii] iv–x, 447 pp. [2 pp. errata]; 17 leaves of plates (all folding); 20 x 12 cm.

Vol. 2, pt. 1: [iii] iv–vi, 312 pp. [1 p. errata]; 18 leaves of plates (all folding); 20 x 12 cm.

Vol. 2, pt. 2: [iii] iv, 284 pp.; 5 leaves of plates (all folding); 20 x 12 cm.

Vol. 3: [iii] iv, 404 pp.; 24 leaves of plates (all folding); 20 x 12 cm.

Comprehensive treatise on hydraulic engineering by a Royal Danish land inspector.

288
MASETTI, GIOVANNI BATTISTA

NOTE ED AGGIUNTE AGLI ELEMENTI DI MECCANICA E D'IDRAULICA DEL PROFESSORE GIUSEPPE VENTUROLI. Bologna: Cardinali e Frulli, 1827.

Vol. 1: [x] 322 [+2] pp.; 3 leaves of plates (all folding); 20 x 12 cm.

Vol. 2: [ii] 3–383 [+1] pp.; 4 leaves of plates (all folding); 20 x 12 cm.

Like item 286, a supplement to the work of Venturoli.

289
MASETTI, GIOVANNI BATTISTA

NOTE ED AGGIUNTE AGLI ELEMENTI DI MECCANICA E D'IDRAULICA DEL PROFESSORE GIUSEPPE VENTUROLI. Napoli: Saverio Starita, 1852.

Vol. 1: [iii] iv–viii, 280 pp.; 21 x 12 cm.

Vol. 2: [ii] 310 [+1] pp.; 7 leaves of plates (all folding); 21 x 12 cm. Two volumes in one.

290
BELANGER, JEAN BAPTISTE CHARLES JOSEPH (1789–1874)

ESSAI SUR LA SOLUTION NUMÉRIQUE DE QUELQUES PROBLÈMES RELATIFS AU MOUVEMENT PERMANENT DES EAUX COURANTES. A Paris: Carilian-Goeury, 1828.

[iii] iv, 38 pp.; 4 leaves of plates; 24 x 16 cm. Photocopy.

Herein Belanger, of the Corps des ponts et chaussées, independently derived the backwater equation and applied it with Prony's resistance formula to several cases of flow. He was less successful in analyzing the hydraulic jump.

291
ALLEN, ZACHARIAH (1795–1882)
THE SCIENCE OF MECHANICS, as applied to the present improvements in the useful arts in Europe, and in the United States of America: adapted as a manual for mechanics and manufacturers, and containing tables and calculations of general practical utility. Providence: Hutchens & Cory, 1829.
[iii] iv–v, 364 pp. [1 p. errata]; illus.; 3 leaves of plates (all folding); 21 x 13 cm.

292
BIGELOW, JACOB (1787–1879)
ELEMENTS OF TECHNOLOGY, taken chiefly from a course of lectures delivered at Cambridge, on the application of the sciences to the useful arts. . . . Boston: Hilliard, Gray, Little, and Wilkins, 1829.
xii, 507 pp.; 22 pp. of plates (some folding); Microfilm.

Published only four months after Allen's shorter, more practical book, the Harvard professor Bigelow's more descriptive work contained less hydraulics but a greater familiarity with European sources.

293
GAUSS, KARL FRIEDRICH (1777–1855)
PRINCIPIA GENERALIA THEORIAE FIGURAE FLUIDORUM IN STATU AEQUILIBRII. [Göttingen]: [s.n.], [1830].
[39]–88 pp.; 27 x 22 cm.

Written in 1829 and published in Commentationes Societatis regiae scientiarum göttingensis recentiores, Vol. VII, 1830.

294
EMY, AMAND ROSE
DU MOUVEMENT DES ONDES ET DES TRAVUAX HYDRAULIQUES MARITIMES. A Paris: Anselin, 1831.
[v] vi–viii, 188 pp. [+2 pp. errata]; 24 x 16 cm. Photocopy.

Comprehensive treatment of wave behavior and application to harbor and coast protection.

295
COCCONCELLI, ANTONIO
ISTITUZIONI D'IDRAULICA TEORICO-PRATICA. Parma: Rossetti, 1832.
Vol. 1: [iii] iv–x, [11]–610 pp. [6 pp. errata]; 6 leaves of plates (all folding); 23 x 15 cm.
Vol. 2: [iii] iv–xxvii, [3]–299 pp. [1 p. errata]; 1 leaf of plates (folding); 23 x 15 cm.

Elements of hydraulic engineering.

296
RENWICK, JAMES (1792–1863)
THE ELEMENTS OF MECHANICS. Philadelphia: Carey & Lea, 1832.
[vii] viii–xxxi [1 p. errata] 508 pp.; illus.; 23 x 12 cm. Microfilm.

A Columbia professor and engineering consultant, Renwick, in this little book, showed a wide familiarity with continental literature.

297
CROZET-FOURNEYRON, MARCEL
INVENTION DE LA TURBINE. Historique suivi d'une note sur un régulateur à mouvement louvoyant applicable aux turbines hydrauliques. Paris et Liége: Ch. Béranger, [1924?]
[ii] 3–55 [+1] pp.; illus.; 1 leaf of plates; 26 x 17 cm.

298
D'AUBUISSON DE VOISINS, JEAN FRANÇOIS (1769–1841)
TRAITÉ D'HYDRAULIQUE, A L'USAGE DES INGÉNIEURS. Paris: F. G. Levrault, 1834.
[v] vi–xii, 564 pp. [+2 pp. errata]; 4 leaves of plates (all folding); 20 x 12 cm.

Widely read and translated and sometimes called the first didactic work on hydraulics, d'Aubuisson's text is probably given more credit than it deserves.

299
D'AUBUISSON DE VOISINS, JEAN FRANÇOIS (1769–1841)
TRAITÉ D'HYDRAULIQUE, À L'USAGE DES INGÉNIEURS. Seconde édition. Paris: Pitois-Levrault et c^{ie}, 1840.
[v] vi–xii, 646 [2] pp.; 5 leaves of plates (all folding); 22 x 13 cm.

300
D'Aubuisson de Voisins, Jean François (1769–1841)
TRATTATO D'IDRAULICA AD USO DEGL'INGEGNERI. . . . [Translated by Filippo Maria Pagano.] Napoli: Presso Serafino Prestia Tipografo, 1844.
[iv] 5–601 pp. [+2 pp. errata]; 5 leaves of plates (all folding); 22 x 14 cm. Taken from the 2nd French edition.

301
D'Aubuisson de Voisins, Jean François (1769–1841)
A TREATISE ON HYDRAULICS, FOR THE USE OF ENGINEERS. Translated from the French by Joseph Bennett. Boston: Little, Brown and Company, 1852.
[iii] iv–xxiii [1 p. errata] 532 pp.; 5 leaves of plates (all folding); 22 x 13 cm.

302
Beaufoy, Mark (1764–1827)
NAUTICAL AND HYDRAULIC EXPERIMENTS, with numerous scientific miscellanies. London: Printed at the private press of Henry Beaufoy, 1834.
Vol. I: [vii] ii–cxix [+i], 688 pp.; frontis.; 16 leaves of plates; 31 x 25 cm. Vols. II and III never published.

Drag tests made by pendulum on many elementary forms of immersed body.

303
Darcy, Henri Philibert Gaspard (1803–1858)
RAPPORT À M. LE MAIRE ET AU CONSEIL MUNICIPAL DE DIJON SUR LES MOYENS DE FOURNIR L'EAU NÉCESSAIRE A CETTE VILLE. Dijon: Douillier, 1834.
[iv] 120 pp.; 1 leaf of plates (folding); 28 x 22 cm.

An able member of the Corps des ponts et chaussées stationed at the city of his birth as inspector general, Darcy contributed notably to knowledge of flow in pipes and through porous media.

304
Darcy, Henri Philibert Gaspard (1803–1858)
RECHERCHES EXPÉRIMENTALES RELATIVES AU MOUVEMENT DE L'EAU DANS LES TUYAUX. Paris: Mallet-Bachelier, 1857.

Vol. 1 (text): [v] vi–xxii, [iv] 268 pp.; 27 x 21 cm.
Vol. 2 (atlas): [iv] and plates; 24 leaves of plates (all double); 43 x 30 cm.

305
DARCY, HENRI PHILIBERT GASPARD (1803–1858) and HENRI ÉMILE BAZIN (1829–1917)
RECHERCHES HYDRAULIQUES, ENTREPRISES PAR M.H. DARCY . . . continuées par M.H. Bazin. . . . Paris: Text, Imprimerie Impériale [Dunod], 1865; Atlas, Dunod, Vol. I, 1865; Vol. II, 1865.
Text, Vol. I: [v] vi–xxxii, 501, vii–xxxii pp.; 27 x 21 cm.
Vol. II: [v] vi–xviii, 152 pp.; 27 x 21 cm. Two volumes in one.
Atlas, Vol. I: [iv] pp.; 34 x 26 cm.
Vol. II; [iv] pp.; 66 leaves of plates (all double); 34 x 26 cm. Two volumes in one. Extracted from vol. 19 of Mémoires présentés pars divers savants à l'Institute Impérial de France.

In ill health, Darcy was assisted in his experimental work on open-channel resistance and wave motion by his protégé in the Corps, Bazin, who brought the experiments to a successful conclusion.

306
MATTHEWS, WILLIAM, engineer.
HYDRAULIA; an historical and descriptive account of the water works of London, and the contrivances for supplying other great cities, in different ages and countries. London: Simpkin, Marshall, and Co., 1835.
[v] vi–xx, 454 pp.; frontis.; 18 leaves of plates (6 folding); 22 x 13 cm.

307
STORROW, CHARLES STORER (1809–1904)
A TREATISE ON WATER-WORKS FOR CONVEYING AND DISTRIBUTING SUPPLIES OF WATER; with tables and examples. Boston: Hilliard, Gray, and Co., 1835.
[v] vi–xii, 242 pp.; illus.; 17 x 10 cm.

Early American résumé of French writings.

308
MORIN, ARTHUR JULES (1795–1880)
EXPÉRIENCES SUR LES ROUES HYDRAULIQUES À AUBES PLANES, et sur les roues hydrauliques à augets. Metz: Mme. Thiel, 1836.

[iii] iv–xv, 133 pp.; 3 leaves of plates (all folding); 27 x 21 cm.

A treatise on water wheels by a professor of machinery at Metz.

309
MORIN, ARTHUR JULES (1795–1880)
EXPÉRIENCES SUR LES ROUES HYDRAULIQUES À AXE VERTICAL APPELÉES TURBINES. Metz: Mme. Thiel, 1838.
[iii] iv, 52 pp.; 2 leaves of plates (both folding); 26 x 21 cm.

310
MORIN, ARTHUR JULES (1795–1880)
HYDRAULIQUE. Deuxième édition. Paris: L. Hachette et cie, 1858.
[iv] 528 pp.; illus.; 7 leaves of plates (all folding); 21 x 13 cm. At head of title: "Leçons de mécanique practique".

A solid textbook on practical hydraulics.

311
MORIN, ARTHUR JULES (1795–1880)
DES MACHINES ET APPAREILS DESTINÉS A L'ÉLÉVATION DES EAUX. Paris: L. Hachette et cie, 1863.
[iv] 323 pp.; illus.; 9 leaves of plates (all folding); 22 x 13 cm.

An illustrated review of all forms of pumping machinery.

312
MEISSNER, WILHELM
NEUE THEORIE DES UFER- UND STROMBAUES, erläutert durch die Geschichte der Wasserbauten an den Süder-Dithmarsichen Elbufer. Hamburg: Hoffman und Campe, 1837.
[iii] iv–xv [i], 230 pp.; 3 leaves of plates (all folding); 20 x 12 cm.

313
RÜHLMANN, MORITZ (1811–1890)
DIE HORIZONTALEN WASSERRAEDER UND BESONDERS DIE TURBINEN ODER KREISELRÄDER, ihre Geschichte, Construction und Theorie. Chemnitz: Expedition des Gewerbeblattes für Sachsen, 1840.
[iii] iv–vi, 52 pp. [1 p. errata]; 4 leaves of plates; 26 x 22 cm. Lacks plates II and III.

A good historical, practical, and theoretical treatment. Compare with Poncelet (item 280).

314
HAGEN, GOTTHILF HEINRICH LUDWIG (1797–1884)
HANDBUCH DER WASSERBAUKUNST. [Vol. 1], Erster Theil: Die Quellen. [Vols. 2 and 3], Zweiter Theil: Die Ströme. [Vol. 4], Die Ströme und Kanäle. Königsberg in Preussen: Gebrüder Bornträger, Vol. 1, 1841; Vol. 2, 1844; Vol. 3, 1847; Vol. 4, 1852.
Vol. 1: [iii] iv–viii, 807 pp.; 21 x 13 cm. Lacks plates.
Vol. 2: [iii] iv, 516 pp.; 21 x 13 cm. Lacks plates.
Vol. 3: [iii] iv, 500 pp.; 21 x 13 cm. Lacks plates.
Vol. 4: [iii] iv, 745 [+3] pp.; 21 x 13 cm. Volumes 2 and 3 bound together. Lacks plates.

Hagen was an able government engineer, teacher, and writer, but paradoxically his most original work (not included in the collection) preceded by some three decades the comparable demonstration of instability for which Osborne Reynolds became famous.

315
HAGEN, GOTTHILF HEINRICH LUDWIG (1797–1884)
HANDBUCH DER WASSERBAUKUNST. [Vols. 1–2], Erster Theil: Die Quellen. Brunnen, Wasserleitungen und Fundirungen. 3rd edition. [Vols. 3–6], Zweiter Theil: Die Ströme. Uferschälungen, Strombauten und Schiffahrts-Canäle. 3rd edition. [Vols. 7–10], Dritter Theil: Das Meer. Seeufer-und Haffen-Bau. [1st edition]. [Vols. 11–15], Atlas. Berlin: Ernst & Korn, Vol. 1, 1869, Vol. 2, 1870; Vol. 3, 1871; Vol. 4, 1873; Vols. 5 and 6, 1874; Vols. 7 and 8, 1863; Vol. 9, 1864; Vol. 10, 1865.
Vol. 1: [xiv] 383 pp.; 21 x 14 cm.
Vol. 2: [vi] 419 pp.; 21 x 14 cm.
Vol. 3: [v] vi, 413 pp.; 21 x 14 cm.
Vol. 4: [v] vi, 399 pp.; 21 x 14 cm.
Vol. 5: [v] vi, 372 pp.; 21 x 14 cm.
Vol. 6: [v] vi, 415 pp.; 21 x 14 cm.
Vol. 7: [v] vi–xviii [ii] 364 pp.; 22 x 14 cm.; Lacks plates.
Vol. 8: [vi] 407 pp.; 22 x 14 cm.; Lacks plates.
Vol. 9: [v] vi, 428 pp.; 22 x 14 cm.; Lacks plates.
Vol. 10: [v] iv, 398 pp.; 22 x 14 cm.; Lacks plates.
Vol. 11: 46 leaves of plates (all double); 28 x 19 cm.
Vol. 12: 22 leaves of plates (all double); 28 x 19 cm.
Vol. 13: 32 leaves of plates (all double); 28 x 19 cm.
Vol. 14: 42 leaves of plates (all double); 28 x 19 cm.
Vol. 15: 20 leaves of plates (all double); 28 x 19 cm.

316
KELVIN, WILLIAM THOMSON, BARON (1824–1907)
MATHEMATICAL AND PHYSICAL PAPERS. Vol. IV: Hydrodynamics and general dynamics. Arranged and revised with brief annotations by Sir Joseph Larmor. Cambridge [England]: [Cambridge] University Press, Series: 1882–1911, Vol. IV: 1910.
[v] vi–xv [+i errata], 563 pp. [+1 p. adv.]; illus.; 22 x 14 cm.

Chancellor of the University of Glasgow and Fellow of St. Peter's College at Cambridge, Kelvin wrote on hydrodynamics (particularly vortex motion), theory of the tides, water waves, and general dynamics.

317
PIOLA, GABRIO (1791–1850)
NUOVE RICERCHE PER UNA RISOLUZIONE PIU RIGOROSA DI VARI PROBLEMI SUL MOTO DELL'ACQUA. . . . [Milan: Tipografia Bernardoni; 1843].
[3] 4–97 pp.; 27 x 22 cm. Originally published in Vol. 1, R. Istituto di scienze e lettere. Memorie. Milan: 1843, pp. 217–311.

A mathematical treatment of two-dimensional, three-dimensional, and free-surface flow.

318
PIOLA, GABRIO (1791–1850)
SUL MOTO PERMANENTE DELL'ACQUA. [Milan: Tipografia Bernardoni; 1845].
[3] 4–92 pp.; 30 x 22 cm. Originally published in Vol. 2, R. Istituto lombardo di scienze e lettere. Memorie. Milan: 1845, pp. 53–152.

319
BARRÉ DE SAINT-VENANT, ADHÉMAR JEAN CLAUDE (1797–1886)
NOTE À JOINDRE AU MEMOIRE SUR LA DYNAMIQUE DES FLUIDES, présenté le 14 avril 1834. [Paris]: Bachelier, [1843].
5 pp.; 27 x 22 cm. "Extrait des Comptes rendus des séances de l'Académie des Sciences, séance du 27 novembre 1843." Vol. 17, 1843 (2) Presentation copy to l'Ecole des ponts et chaussées.

Saint-Venant was an extremely able writer on hydraulics as well as on his primary love, elasticity, publishing papers on wave motion, resistance, and gradually varied flow. This note is a terse generalization of the Navier-Stokes (and Reynolds) equations of acceleration.

320
BARRÉ DE SAINT-VENANT, ADHÉMAR JEAN CLAUDE (1797–1886)
MÉMOIRE SUR UN MODE D'INTERPOLATION APPLICABLE À DES QUESTIONS RELATIVES AU MOUVEMENT DES EAUX, et suppléant à l'intégration souvent impossible des équations aux dérivées partielles. [Paris]: Bachelier, [1843?].
8 pp.; 28 x 22 cm. "Extrait des 'Comptes rendus des séances de L'Académie des Sciences', séance du 13 novembre 1843." Vol. 17, 1843, (2).

321
BARRÉ DE SAINT-VENANT, ADHÉMAR JEAN CLAUDE (1797–1886) and ______ WANTZEL
NOUVELLES EXPÉRIENCES SUR L'ÉCOULEMENT DE L'AIR DÉTERMINÉ PAR DES DIFFÉRENCES DE PRESSION CONSIDÉRABLES. [Paris]: Bachelier, [1843?].
3 pp.; 27 x 22 cm. "Extrait des 'Comptes rendus des séances de l'Académie des Sciences,' séance du 13 novembre 1843." [Vol. 17, 1843, (2)]

322
BARRÉ DE SAINT-VENANT, ADHÉMAR JEAN CLAUDE (1797–1886)
DES DIVERSES MANIÈRES DE POSER LES ÉQUATIONS DU MOUVEMENT VARIÉ DES EAUX COURANTES. Paris: Ch. Dunod, 1887.
148–228 pp.; 20 x 13 cm. This essay was written in 1872, but published in 1887 in the Annales des ponts et chaussées. Mémoires et Documents. 6e série, Tome XIII. It is item No. 12 in that volume. This copy was apparently removed from that volume.

323
BOUSSINESQ, JOSEPH (1842–1929) and ALFRED AIMÉ FLAMANT (*b.* 1839)
NOTICE SUR LA VIE ET LES TRAVAUX DE BARRÉ DE SAINT-VENANT. Paris: Ch. Dunod, 1886.
39 pp.; 23 x 14 cm. "Extrait des 'Annales des ponts et chaussées,' novembre 1886."

324
[AIRY, SIR GEORGE BIDDELL] (1801–1892)
TIDES AND WAVES. [London] [B. Fellowes] [1845?]
241–396 pp.; 27 x 21 cm. Extracted from the Encyclopaedia Metropolitana.

Director of the Cambridge Observatory and later Astronomer Royal at Greenwich, Airy wrote as an encyclopedia article what is probably the best-known treatise on wave motion.

325
REDTENBACHER, FERDINAND JACOB (1809–1863)
THEORIE UND BAU DER WASSERRÄDER. Mannheim: Friedrich Bassermann, 1846.
[iii] ii–xii, 324 pp.; 6 leaves of plates; 23 x 15 cm.

Professor of mechanical engineering at Carlsruhe.

326
DUPUIT, JULES (1804–1866)
ETUDES THÉORIQUES ET PRATIQUES SUR LE MOUVEMENT DES EAUX COURANTES, suivies de considérations relatives au régime des grandes eaux, au débouché à leur donner, et à la marche des alluvions dans les rivières à fond mobile. Paris: Carilian-Goeury et V^or Dalmont, 1848.
[v] ii–xii, 275 pp.; 2 leaves of plates (both folding); 23 x 15 cm.

One of the many research-minded members of the Corps des ponts et chaussées, Dupuit wrote extensively on open-channel flow, including both gradually varied flow and flood waves.

327
DUPUIT, JULES (1804–1866)
ÉTUDES THÉORIQUES ET PRATIQUES SUR LE MOUVEMENT DES EAUX DANS LES CANAUX DÉCOUVERTS ET A TRAVERS LES TERRAINS PERMÉABLES[,] avec des considérations relatives au régime des grandes eaux, au déboucher à leur donner, et à la marche des alluvions dans les rivières à fond mobile. Deuxieme édition, revue et considérablement augmentée. Paris: Dunod, 1863.
[v] vi–xxviii, 304 pp.; 6 leaves of plates (all folding); 28 x 22 cm. This is the 2nd edition of item 326.

328
DUPUIT, JULES (1804–1866)
TRAITÉ THÉORIQUE ET PRATIQUE DE LA CONDUITE ET DE LA DISTRIBUTION DES EAUX . . . suivi d'un extrait de l'essai[,] Sur les moyens de conduire, d'élever et de distribuer les eaux[,] par Genieys . . . et de la description des filtres naturels de Toulouse par D'Aubuisson. . . . Paris: Carilian-Goeury et V^or Dalmont, 1854.

Text, Vol. 1: [v] vi–xxiv, 361, [2], 164 pp.; 27 x 20 cm.
Atlas, Vol. 2: [iv] pp.; 96 leaves of plates; 32 x 25 cm.

329
DUPUIT, JULES (1804–1866)
TRAITÉ THÉORIQUE ET PRATIQUE DE LA CONDUITE ET DE LA DISTRIBUTION DES EAUX. Deuxième édition, revue et considérablement augmentée. Paris: Dunod, 1865.
Vol. 1 (text): [v] vi–xv, 496 pp.; 27 x 21 cm.
Vol. 2 (atlas): [iv] pp.; 94 leaves of plates (all double); 33 x 25 cm.

330
BARLOW, PETER (1776–1862) and JOHN F. W. HERSCHEL (1792–1871)
THE ENCYCLOPAEDIA OF MECHANICAL PHILOSOPHY: comprising the sciences of mechanics, hydrodynamics, pneumatics, and optics. London: John Joseph Griffin and Company, 1848.
[v] vi–viii, 586 pp.; 78 leaves of plates; 29 x 22 cm. This work forms part of volumes 3 and 4 of the Encyclopaedia Metropolitana.

A veritable "dictionary of knowledge" with some 230 pages devoted to hydraulics and pneumatics.

331
WEISBACH, JULIUS (1806–1871)
PRINCIPLES OF THE MECHANICS OF MACHINERY AND ENGINEERING. [Translated by "J. R."] London: Hippolyte Bailliere, Vol. 1, 1847; Vol. 2, 1848.
Vol. 1: [v] vi–xvii, 551 pp. [+8 pp. adv.]; illus.; 21 x 14 cm.
Vol. 2: [5] viii–xv, 420, 4 pp., 7 pp. adv., [9] pp.; illus.; frontis.; 21 x 14 cm. (Library of Illustrated Standard Scientific Works, Vol. II and Vol. V)

Weisbach, a teacher of applied mathematics at the Freiberg school of mines, deserved the popularity accorded d'Aubuisson, for his books were original, modern, and attractively illustrated. Among other innovations, his is the pipe-resistance relationship presently used. The original German edition of his textbook was published in 1845.

332
WEISBACH, JULIUS (1806–1871)
PRINCIPLES OF THE MECHANICS OF MACHINERY AND ENGINEERING. First American edition. Philadelphia: Lea and Blanchard, 1848.

Fig. 230.

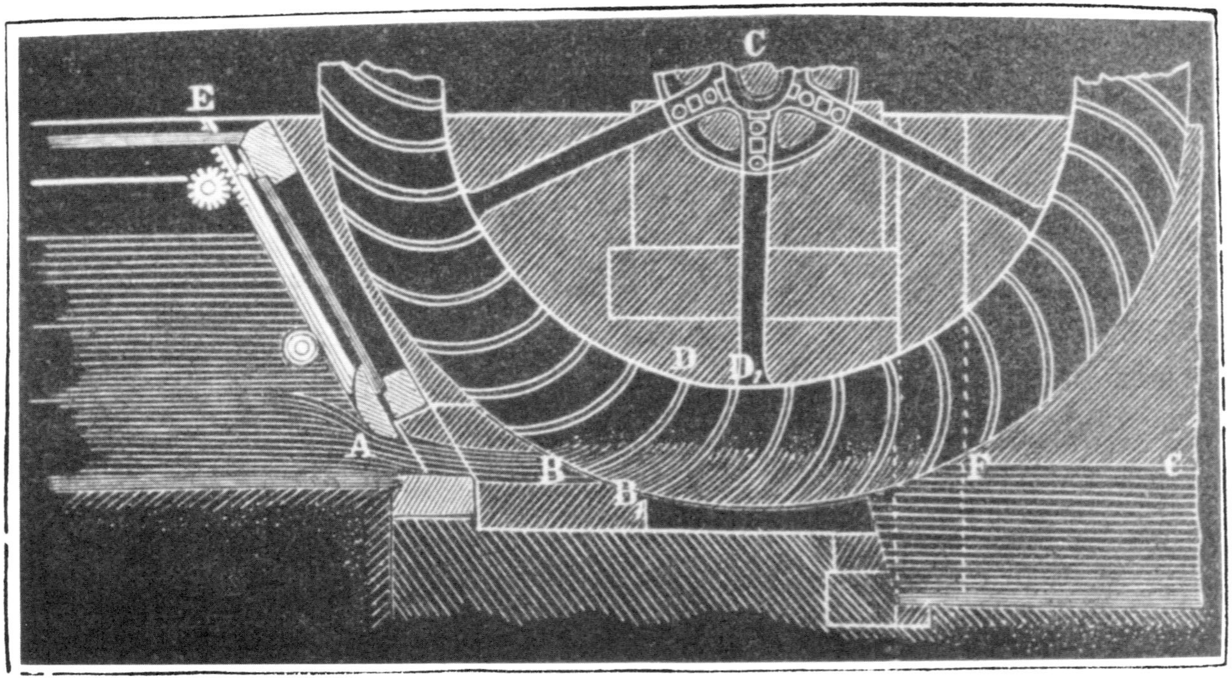

332. Weisbach's illustration of a Poncelet water wheel.

Vol. 1: [iii] iv–xv, [25]–486 pp. [+1 p. errata]; illus.; 23 x 15 cm.
Vol. 2: [iii] vi–xvi, [14]–368 pp. [+16 pp. adv.]; illus.; frontis.; 23 x 15 cm.

333
Weisbach, Julius (1806–1871)
LEHRBUCH DER INGENIEUR- UND MASCHINEN-MECHANIK. 2nd edition. Braunschweig: Friedrich Vieweg und Sohn, Vol. 1, 1850; Vols. 2 & 3, 1851.
Vol. 1: [4 pp. adv.] [v] vi–xiv, 696 pp. [+1 p. errata]; illus.; 22 x 13 cm.
Vol. 2: [7] 8–712, viii pp.; illus.; 22 x 13 cm.
Vol. 3: [7] 8–192 pp. [vi pp. adv.]; illus.; 22 x 13 cm. Three volumes in two.

334
Weisbach, Julius (1806–1871)
LEHRBUCH DER INGENIEUR-UND MASCHINEN-MECHANIK. Erster Theil: Theoretische Mechanik. 5th edition. Edited by Gustav Herrmann. Braunschweig: Friedrich Vieweg und Sohn, 1875.
[v] ii–xii, 1311 pp.; illus.; 22 x 14 cm.

335
Undeutsch, Hermann (1844–1912)
ZUM GEDÄCHTNIS AN OBERBERGRAT PROFESSOR DR. H. C. JULIUS LUDWIG WEISBACH ANLÄSSLICH SEINER HUNDERTJÄHRIGEN GEBURTSTAGSFEIER. Freiberg i.S.: Craz & Gerlach, 1906.
[3] 4–15 pp.; frontis.; 23 x 16 cm. Photocopy.

336
JULIUS WEISBACH. GEDENKSCHRIFT ZU SEINEM 150. GEBURTSTAG. Edited and collected together by Werner Beck. Berlin: Akademie Verlag, 1956.
[11] 12–173 [+3] pp.; illus.; frontis.; 1 leaf of plates (folding); 24 x 15 cm. (Freiberger Forschungshefte. Kultur und Technik, D 16.) Photocopy.

337
Delaunay, Charles Eugene (1816–1872)
COURS ÉLÉMENTAIRE DE MÉCANIQUE THÉORIQUE ET APPLIQUÉE. Paris: Langlois & Leclercq, 1851.
[iv], 696 pp.; illus.; 17 x 11 cm.

Profusely illustrated treatise by a professor of the Parisian Faculté des Sciènces.

338
EWBANK, THOMAS (1792–1870)
A DESCRIPTIVE AND HISTORICAL ACCOUNT OF HYDRAULIC AND OTHER MACHINES FOR RAISING WATER, ANCIENT AND MODERN; with observations on various subjects connected with the mechanic arts, including the progressive development of the steam engine. . . . Twelfth edition. New York: Bangs, Platt, & Co., 1851.
[iii] iv–xvi, 608 pp.; illus.; 22 x 14 cm.

Well illustrated but as garrulous as the title.

339
STOKES, SIR GEORGE GABRIEL (1819–1903)
ON THE EFFECT OF THE INTERNAL FRICTION OF FLUIDS ON THE MOTION OF PENDULUMS. Cambridge [England]: "Printed at the Pitt Press" [University Press], 1851.
[ii], 99 pp.; 27 x 20 cm. Transactions of the Cambridge Philosophical Society, Vol. IX, Part II.

Stokes was the last of four scientists to derive the general equations of fluid acceleration, but probably the first (after at least one false start) to apply them correctly.

340
STOKES, SIR GEORGE GABRIEL (1819–1903)
MATHEMATICAL AND PHYSICAL PAPERS. Cambridge [England]: The University Press, Vol. 1, 1880; Vol. 2, 1883.
Vol. 1: x, 328 pp., +31 pp. adv.; 22 x 14 cm.
Vol. 2: viii, 366 pp., +31 pp. adv.; 22 x 14 cm.

A collection of some 35 early papers, particularly on the internal friction of fluids and the theory of oscillatory waves.

341
STOKES, SIR GEORGE GABRIEL (1819–1903)
MATHEMATICAL AND PHYSICAL PAPERS. [Vols. 1–3 edited by the author; Vols. 4&5 edited by J. Larmor] Cambridge [England]: [Cambridge] University Press, Vol. 1, 1880; Vol. 2, 1883; Vol. 3, 1922; Vol. 4, 1904; Vol. 5, 1905.

Vol. 1: [v] vi–x, 328 pp.; 21 x 14 cm.
Vol. 2: [v] viii, 366 pp.; 21 x 14 cm.
Vol. 3: [vii] viii, 415 pp.; 1 leaf of plates; 21 x 14 cm. Reprint of 1901 edition.
Vol. 4: [v] vi–viii, 378 pp.; illus.; frontis.; 21 x 14 cm.
Vol. 5: [v] vi–xxv, 370 pp.; frontis.; 21 x 14 cm.

Vols. 1 and 2, a collection of some 35 early papers, particularly on the internal friction of fluids and the theory of oscillatory waves; Vol. 3, additional early papers, particularly on the motion of pendulums; Vol. 4, papers from 1853 to 1876, but little on fluids; Vol. 5, obituary by Lord Rayleigh and remaining papers, a few on fluids.

342
BOURNE, JOHN, civil engineer
A TREATISE ON THE SCREW PROPELLER, WITH VARIOUS SUGGESTIONS OF IMPROVEMENT. London: Longman, Brown, Green and Longmans, 1852.
[v] iv–viii [iv], 243, i–xliv pp.; illus.; 20 leaves of plates (11 folding); 17 x 20 cm. With 3 manuscript leaves tipped-in, unsigned.

Comprehensive account of the history, design, and operation of ship propellers.

343
LARDNER, DIONYSIUS (1793–1859)
A TREATISE ON HYDROSTATICS AND PNEUMATICS. New edition revised and corrected. London: Longmans, Brown, Green & Longmans, 1852.
[iii] vi–viii, 361 pp.; illus.; 16 x 10 cm.

344
ELLET, CHARLES (1810–1862)
THE MISSISSIPPI AND OHIO RIVERS: CONTAINING PLANS FOR THE PROTECTION OF THE DELTA FROM INUNDATION; and investigations of the practicability and cost of improving the navigation of the Ohio and other rivers by means of reservoirs. Philadelphia: Lippincott, Grambo, and Co., 1853.
[v] vi–viii, [17]–367 pp.; 12 leaves of plates (2 folding); 25 x 16 cm.

This is noteworthy because of the Corps of Engineers' opposition to such methods of flood control.

345
GUMPPENBERG-PÖTTMES, JOSEPH, FREIHERR VON

DER WASSERBAU AN GEBIRGSFLÜSSEN; oder Praktische Anleitung, die Gebirgsflüsse dort, wo sie in zerstreuten Rinnsalen das flache Land verwüsten, und endlich die sie aufnehmenden Hauptströme versanden, ihrer Natur entsprechend zu behandeln. Augsberg: Matth. Rieger, 1854.

[iii] iv–viii, 83 pp.; 5 leaves of plates (all folding); 28 x 22 cm.

346
DELPRAT, ISAAC PAUL (1793–1880)

OVER DE BETREKKING TUSSCHEN DE GEMIDDELDE SNELHEID VAN HET WATER IN RIVIEREN EN DE WERKELIJKE SNELHEID. Amsterdam: C.G. Van der Post, 1855.

[ii] 71 pp.; 9 leaves of plates, 28 x 21 cm.

347
FRANCIS, JAMES BICHENO (1815–1892)

LOWELL HYDRAULIC EXPERIMENTS. . . . Boston: Little, Brown, and Company, 1855.

[v] vi–xi, 156 pp.; 30 leaves of plates (all folding); 32 x 25 cm.

The popularity of this tome led to its author's name being applied to the mixed-flow type of turbine, even though his own wheels were strictly of the radial-flow category.

348
FRANCIS, JAMES BICHENO (1815–1892)

LOWELL HYDRAULIC EXPERIMENTS. . . . Fourth edition. New York: D. Von Nostrand, 1883.

[v] vi–xi, 286 pp.; 46 leaves of plates (all double); 32 x 23 cm.

349
MÜLLER, FRANZ

DIE GEBIRGS-BÄCHE UND IHRE VERHEERUNGEN WIE DIE MITTEL ZUR ABWENDUNG DER LETZTEREN. Landshut: J. O. Wolfle [Krüll'sche Universitäts-Buchhandlung], 1857.

[v] vi–viii, 49 pp.; 6 leaves of plates (all folding); 23 x 15 cm.

350
ARMENGAUD, JACQUES EUGÈNE (1810–1891)

TRAITÉ THÉORIQUE ET PRATIQUE DES MOTEURS HYDRAULIQUES. Nouvelle édition. Paris: chez l'auteur, 1858.

Vol. 1 (text): [v] vi–viii, 500 pp.; illus.; 27 x 21 cm.
Vol. 2 (atlas): [iii] pp.; 42 leaves of plates (all double); 27 x 21 cm.

An extensive, well-illustrated text on hydraulic turbines by a former professor at the Conservatoire Impériale des Arts et Métiers.

351
MÉNABRÉA, LUIGI FEDERIGO (1809–1896)
NOTE SUR LES EFFETS DU CHOC DE L'EAU DANS LES CONDUITES. [Paris: Mallet-Bachelier, 1858?].
4 pp.; 27 x 22 cm. "Extrait des Comptes rendus des séances de l'Académie des Sciences, tome XLVII, séance du 2 Août 1858."

352
RACHMANINOW, ______
NOTE SUR LA THÉORIE DE LA ROUE HYDRAULIQUE EN DESSOUS À AUBES PLANES. [Paris: Mallet-Bachelier, 1858?].
21 pp.; 28 x 22 cm. "Extrait du Journal de Mathématiques pures et appliquées, 2e série, tome III, 1858."

353
BRESSE, JACQUES ANTOINE CHARLES (1822–1883)
COURS DE MÉCANIQUE APPLIQÚEE, professé à l'Ecole impériale des ponts et chaussées. Première partie [Vol. 1]: Résistance des matériaux et stabilité des constructions. Troisiême partie [Vol. 3]: Calcul des moments de flexion dans une poutre à plusieurs travées solidaires. Paris: Vol. 1, Mallet-Bachelier, 1859; Vol. 3, Gauthier-Villars, 1865.
Vol. 1: [v] vi–xxiv, 470 pp. [+1 p. errata]; illus.; 21 x 13 cm.
Vol. 3: [v] vi–xxvi, [1 p. errata] 360 pp.; illus.; 1 leaf of plates (folding); 21 x 13 cm. Published in three parts; Vol. 2 entitled, Hydraulique. (See item 354).

Successor to Belanger as professor at the Ecole des ponts et chaussées, Bresse correctly expressed the momentum characteristics of the hydraulic jump, integrated the equations of gradually varied flow, and prepared tables for their application. Only the second of the three volumes was devoted to hydraulics.

354
BRESSE, JACQUES ANTOINE CHARLES (1822–1883)
COURS DE MÉCANIQUE APPLIQUÉE, professé à l'Ecole impériale des ponts et chaussées. Second partie [Vol. 2]: Hydraulique. Deuxième édition. Paris: Gauthier Villars, 1868.

[v] vi–xxi [+i p. errata], 568 pp.; illus.; 1 leaf of plates (folding); 21 x 12 cm. The second edition was published in two volumes; Vol. 1 entitled Résistance des matériaux et stabilite des constructions.

Awarded the Poncelet prize by the Académie des sciences in 1874.

355

BRESSE, JACQUES ANTOINE CHARLES (1822–1883)

COURS DE MÉCANIQUE APPLIQUÉE, professé à l'Ecole des ponts et chaussées. Seconde partie. Hydraulique. Troisième édition. Paris: Gauthier-Villars, 1879.

[v] vi–xxi [+i p. errata], 602 pp.; illus.; 1 leaf of plates; 21 x 13 cm. Published in two parts. Part One entitled "Résistance des matériaux et stabilité des constructions."

356

NEVILLE, JOHN

HYDRAULIC TABLES, COEFFICIENTS, AND FORMULAE, FOR FINDING THE DISCHARGE OF WATER FROM ORIFICES, NOTCHES, WEIRS, PIPES AND RIVERS. Second edition. London: John Weale, 1860–61.

[v] ii–xii, 455 pp. [+1–12, 1–32 pp. adv.]; illus.; 22 x 14 cm.

Typical 19th-century practice.

357

HUMPHREYS, ANDREW ATKINSON (1810–1883) and H. L. ABBOT (1831–1927)

REPORT UPON THE PHYSICS AND HYDRAULICS OF THE MISSISSIPPI RIVER; upon the protection of the alluvial region against overflow; and upon the deepening of the mouths. . . . Philadelphia: J. B. Lippincott & Co., 1861.

[iv] v–xxiv, 17–456, cxlvi pp.; 20 leaves of plates; 32 x 24 cm. (Professional papers of the Corps of Topographical Engineers, United States Army. . . . No. 4).

One of the best-known treatises on the Mississippi, the volume contains a wealth of relevant information but less satisfactory recommendations for river improvement.

358

LERSCH, BERNHARD MAXIMILIAN (1817–1902)

HYDRO-PHYSIK; ODER LEHRE VOM PHYSIKALISCHEN VERHALTEN DER NATÜRLICHEN WÄSSER NAMENTLICH

VON DER BILDUNG DER KALTEN UND WARMEN QUELLEN. 2. Aufl. Berlin: August Hirschwald, 1865.
[v] iv, 283 pp.; illus., frontis.; 4 leaves of plates; 22 x 24 cm.

An illustrated work on the geologic aspects of hydraulics—or vice versa.

359
NEUSTADT, CAMILLE
DE L'EMPLOI DE L'EAU COMME MOYEN DE TRANSMISSION DE FORCE DANS LES DOCKS COMMERCIAUX ET LES ARSENAUX MARITIMES. Paris: A. Chaix et cie, 1866.
[5] 6–30 pp.; 2 leaves of plates (both folding); 23 x 15 cm.

360
FOURNIE, VICTOR (*b.* 1837)
RÉSUMÉ DES EXPÉRIENCES HYDRAULIQUES EXÉCUTÉES PAR LE GOUVERNEMENT AMÉRICAIN SUR LE MISSISSIPI [sic] et remarques sur les conséquences qui en découlent relativement à la théorie des eaux courantes. Paris: Dunod, 1867.
[vii] viii–xii, 128 pp.; 28 x 23 cm.

A critical review of item 357.

361
GAUCKLER, PHILIPPE-GASPARD (*b.* 1826)
ÉTUDE THÉORIQUES ET PRATIQUES SUR L'ÉCOULEMENT ET LE MOUVEMENT DES EAUX. [Paris: Gauthier-Villars 1867].
5 pp.; 28 x 22 cm.

French originator of what came to be called the Manning formula.

362
ENGLAND. COMMISSIONERS OF PATENTS
PATENTS FOR INVENTIONS. Abridgements of specifications relating to hydraulics. A.D. 1617–1865. London: George E. Eyre and William Spottiswoode, 1868.
[iii] iv–xlv, 1440 [+8] pp.; 18 x 12 cm.

A compilation of several thousand patent claims.

363
NEUMANN, FRIEDRICH, engineer
HYDRAULISCHE MOTOREN. Bau und Anlage der wichtigsten

vom Wasser getriebenen Maschinen, dargestellt durch Zeichnungen mit Beschreibung und Berechnung von Turbinen, Wasserrädern und Wassersäulenmaschinen. . . . Weimar: Berhard Friedrich Voigt, 1868.

[v] vi–x, 226 pp.; illus.; 21 x 23 cm. (Neuer Schauplatz der Künste und Handwerke, vol. 286).

Both qualitative and quantitative treatment of hydraulic machines.

364
RAYLEIGH, JOHN WILLIAM STRUTT, BARON (1842–1919)

SCIENTIFIC PAPERS. New York: Dover Publications, Inc., [c1964].

Vols. 1 and 2: [v] vi–xiii, 562 pp.; [iii] iv–x, 598 pp.; illus.; 16 pp. of plates; 23 x 15 cm.

Vols. 3 and 4: [ix] x–xiv, 596 pp.; [vii] viii–xiv, 568 pp.; illus.; frontis.; 5 leaves of plates; 23 x 15 cm.

Vols. 5 and 6: [vii] x–xvi, 624 pp.; [v] viii–xiii, 718 pp.; illus.; 1 leaf of plates; 23 x 15 cm.

IaU copy bound in 3 volumes.

A professor at Cambridge and eventually president of the Royal Society, Rayleigh was a prolific writer in many fields. In that of fluid motion he contributed to the mechanics of bubble collapse, wave theory, jet instability, the flexure-heat-laminar flow analogy, and dynamical similarity.

365
REYNOLDS, OSBORNE (1842–1912)

PAPERS ON MECHANICAL AND PHYSICAL SUBJECTS. Vol. 1, 1869–1882; Vol. 2, 1881–1900; Vol. 3, The sub-mechanics of the universe. Cambridge [England]: [Cambridge] University Press, Vol. 1, 1900; Vol. 2, 1901; Vol. 3, 1903.

Vol. 1: [v] viii–xv, 416 pp.; illus.; 2 leaves of plates; 26 x 17 cm.

Vol. 2: [vii] viii–xii, 740 pp.; illus.; 6 leaves of plates; 26 x 17 cm.

Vol. 3: [vii] viii–xvii, 254 pp.; illus.; 26 x 17 cm. Errata slip inserted.

In his 37 years as professor at Manchester, Reynolds wrote on a wide variety of subjects, including stability, turbulence, cavitation, scale-model simulation, and different aspects of thermodynamics (However, see item 314).

366
GIBSON, ARNOLD HARTLEY (1878–1959)

OSBORNE REYNOLDS AND HIS WORK IN HYDRAULICS AND

HYDRODYNAMICS. London: published for the British Council by Longmans, Green and Co., 1948.

[iv], 33 pp.; illus.; frontis.; 2 pp. of plates, 1 leaf of plates (folding); 22 x 15 cm.

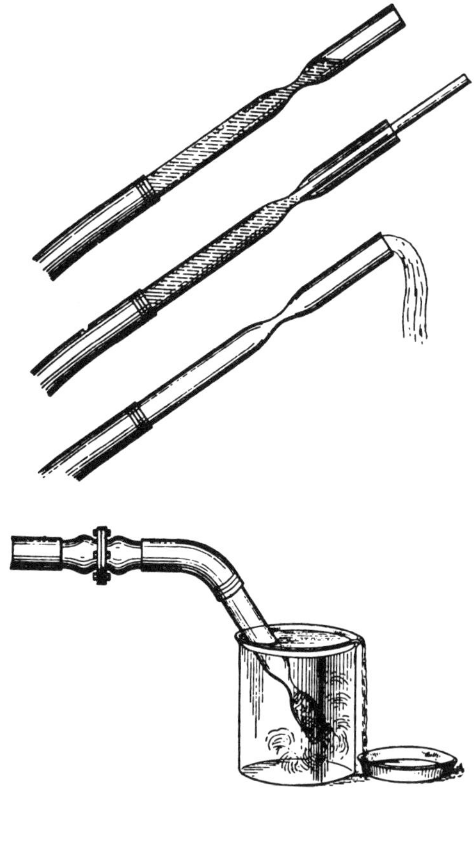

365. Demonstration by Reynolds of cavitation in a glass tube.

367
LOMBARDINI, ELIA (1794–1878)
GUIDA ALLO STUDIO DELL'IDROLOGIA FLUVIALE E DELL'IDRAULICA PRACTICA. Milano: Tipog. e Litog. degli Ingegneri, 1870.
[3] 4–218 pp. [+1 p. errata]; 23 x 16 cm.

An early study in the relatively new field of hydrology.

Fig. 13.

365(2). Reynolds's apparatus for demonstrating instability.

368
LOMBARDINI, ELIA (1794–1878)

DELL'ORIGINE E DEL PROGRESSO DELLA SCIENZA IDRAULICA NEL MILANESE ED IN ALTRE PARTI D'ITALIA. Osservazioni storico-critiche concernenti principalmente i lavori di Leonardo da Vinci, di Benedetto Castelli e di Gian Domenico Guglielmini. Edizione III. . . . Milano, Premiata Lit. E. Tip. degli Ingegneri. Editore B. Saldini, 1872.

[vii] viii–xii, 78 pp.; frontis.; 30 x 21 cm.

369
KLEITZ, CHARLES (1808–1886)

ETUDES SUR LES FORCES MOLÉCULAIRES DANS LES LIQUIDES EN MOUVEMENT ET APPLICATION À L'HYDRODYNAMIQUE. Paris: Dunod, 1873.

[v] vi–xv [+i errata], 240 pp.; illus.; 28 x 22 cm.

A treatise on viscous effects by an inspector general of the Corps des Ponts et Chaussées, including assessment by a committee of which Saint Venant was a member.

370
FROUDE, WILLIAM (1810–1879)

THE PAPERS OF WILLIAM FROUDE. Memoir by Sir Westcott Abell; Evaluation of Froude's work by R.W.L. Gawn. Collected by the Secretary, Captain A. D. Duckworth. London: The Institution of Naval Architects, 1955.

[iv] v–xxii, 343 [+1], 345–358 [+1] pp.; illus.; frontis.; 10 pp. of plates (4 folding), 3 leaves of plates (2 folding); 28 x 21 cm.

One of the prominent names in hydraulic research, Froude established methods of evaluating the results of towing-tank tests on model ships that are still followed in naval establishments. Although the criterion for gravitational similarity that bears his name was formulated by the Frenchman Reech, at least a quarter century earlier, Froude's own contributions were still outstanding.

371
DOWNING, SAMUEL

ELEMENTS OF PRACTICAL HYDRAULICS, FOR THE USE OF STUDENTS IN ENGINEERING AND ARCHITECTURE. Part I. Third edition, revised and enlarged. London: Longmans, Green, and Co., 1875.

[iii] ii, 232 pp. [+48 pp. adv.]; illus.; 22 x 14 cm. Errata slip inserted.

A standard text by a professor at the University of Dublin.

372
Boileau, Pierre Prosper (1811–1891)
NOTIONS NOUVELLES D'HYDRAULIQUE CONCERNANT PRINCIPALEMENT LES TUYAUX DE CONDUITE, LES CANAUX ET LES RIVIÈRES, accompagnées d'une théorie de l'évaluation du travail intermoléculaire des systèmes matériels. Cherbourg: Bedelfontaine et Styffert, 1877.

[5] 6–182 pp; 1 leaf of plates; 23 x 15 cm. Extrait des Mémoires de la Société nationale des sciences naturelles de Cherbourg, tome XXI, 1877.

373
Boussinesq, Joseph (1842–1929)
ESSAI SUR LA THÉORIE DES EAUX COURANTES. [Paris: Imprimerie Nationale 1877]

xxii, 680, 64 pp.; 27 x 21 cm. (Académie des sciences, Paris. Mémoires présentés par divers savants. Sciences mathématiques et physiques. Series 2, Vol. 23, No. 1, Vol. 24, No. 2)

This extensive work is one that still stands as a trustworthy guide to the mechanics of open-channel flow. Though a better writer than teacher, Boussinesq eventually became a professor at the Sorbonne.

374
Fanning, John Thomas (1837–1911)
A PRACTICAL TREATISE ON WATER-SUPPLY ENGINEERING. . . . New York: D. Van Nostrand, 1877.

[v] vi–xxi [iv], 26–619 pp. [+44 pp. adv.]; illus.; frontis.; 32 leaves of plates; 23 x 14 cm. Half-title: Hydraulic engineering and manual for water supply engineers.

A popular engineering text, this work was to go through many editions.

375
Fanning, John Thomas (1837–1911)
A PRACTICAL TREATISE ON HYDRAULIC AND WATER-SUPPLY ENGINEERING . . . Eleventh edition. New York: D. Van Nostrand Company, 1893.

[v] vi–xxii [+iii], 26–644 pp.; illus.; frontis.; 54 leaves of plates; 23 x

15 cm. Half-title: Hydraulic engineering and manual for water supply engineers.

376
NAZZANI, ILDEBRANDO (*b.*1846)
FORMOLE EMPIRICHE PER L'IDRAULICA SPERIMENTALE; NUOVE FORMOLE PER LE PORTATE DEL PÒ E DEL TEVERE. Palermo: Luigi Pedone Lauriel, 1877.
[5], 6–110 pp.; 23 x 15 cm.

377
EMERSON, JAMES, OF WILLIMANSETT, MASS.
TREATISE RELATIVE TO THE TESTING OF WATER-WHEELS AND MACHINERY, with various other matters pertaining to hydraulics. Second edition. Springfield, Mass.: Weaver, Shipman & Company, Printers, 1878.
[6 testimonials & errata, 5] 8–216 pp. [+2 pp. adv.]; illus.; 18 x 11 cm.

378
EMERSON, JAMES, OF WILLIAMANSETT, MASS.
TREATISE RELATIVE TO THE TESTING OF THE WATER-WHEELS AND MACHINERY, also of inventions, studies, and experiments, with suggestions from a life's experience. Sixth edition. [Willimansett, Mass.?]: [s.n.] 1894.
[4] 5–569 pp.; illus.; 18 x 12 cm.

The "suggestions" comprise a most astounding miscellany ranging from myth to morality.

379
LAMB, HORACE (1849–1934)
A TREATISE ON THE MATHEMATICAL THEORY OF THE MOTION OF FLUIDS. Cambridge [England]: [Cambridge] University Press, 1879.
[v] vi–x, 258 pp. [+30 pp. adv.]; illus.; 22 x 14 cm.

Though his original post was at Adelaide, Lamb later became professor at Manchester, where he remained until his retirement in 1920. His primary work was so well received over a century ago that, under the title *Hydrodynamics,* it has been continually revised, enlarged, and republished, remaining in print even to the present day.

380
LAMB, HORACE (1849–1934)
HYDRODYNAMICS. Third edition. Cambridge [England]: [Cambridge] University Press, 1906.
[vii] viii–xv [+i additions and corrections], 634 pp.; illus.; 26 x 16 cm.

381
AUERBACH, FELIX (1856–1933)
DIE THEORETISCHE HYDRODYNAMIK. Nach dem Gange ihrer Entwickelung in der neuesten Zeit in Kürze dargestellt. Braunschweig: Friedrich Vieweg und Sohn, 1881.
[vii] viii–xxi, 150 pp.; illus.; 22 x 14 cm.

382
CUNNINGHAM, ALLAN JOSEPH CHAMPNEYS (1842–1928)
ROORKEE HYDRAULIC EXPERIMENTS. [Vol. 1] Roorkee [India]: Thomason College Press, 1881.
Vol. 1: [iii] iv–xvi, 399 pp.; 24 x 14 cm. Lack Vols. 2 and 3.

A highly detailed account of velocity and discharge measurements at this Indian laboratory.

383
JANICKI, STANISLAW (1872–1939) and L. JACQUET and A. PASQUEAU (*b*.1840?)
IMPROVEMENT OF NON-TIDAL RIVERS. Translated by Bvt. Colonel Wm. E. Merrill. Washington, [D.C.]: Government Printing Office, 1881.
[ii] 194 pp.; illus.; 8 leaves of plates (6 folding); 29 x 23 cm.; two errata slips tipped in.

384
EADS, JAMES BUCHANAN (1820–1887)
PHYSICS AND HYDRAULICS OF THE MISSISSIPPI RIVER. Report of U.S. Levee Commission reviewed by James B. Eads, C.E., St. Louis, February 19, 1876. New Orleans, La.: Picayune Steam Job Print, 1876.
33, 11 pp.; 1 leaf of plates; 23 x 14 cm. Bound with: U.S. Congress. Senate. Committee on Commerce. Memorial of John Cowdon, of Memphis, Tennessee, on the subject of the improvement of the navigation of the Mississippi River and the reclamation of the low lands of

the Mississippi Valley. S. Misc. Report 45–89, 45th Congress, 2nd Session, 1878.

Ead's writings typified the controversy between civilian and military engineers. Noteworthy among the latter was Humphreys (see item 357) who eventually became Chief of Engineers.

385
EADS, JAMES BUCHANAN (1820–1887)
MINORITY REPORT OF MISSISSIPPI RIVER COMMISSION. Washington, D.C.: Gibson Brothers, Printers, 1882.
[ii] 21 pp.; illus.; 23 x 14 cm.

386
GRAEFF, MICHEL IGNACE AUGUSTE (1812–1884)
TRAITÉ D'HYDRAULIQUE, précédé d'une introduction sur les principes généraux de la mécanique. Paris: Imprimerie Nationale, Vol. 1, 1882; Vols. 2 and 3, 1883.
Vol. 1: [v] ii–vii, 329 pp.; illus.; 30 x 23 cm.
Vol. 2: [iv], 537 pp.; illus.; 30 x 23 cm.
Vol. 3: [iv], 48 pp.; illus.; 29 leaves of plates (12 double, 1 double folding); 30 x 23 cm.

An extensive and beautifully illustrated treatise on mechanics, hydrodynamics and hydraulics.

387
CALIGNY, ANATOLE FRANÇOIS HÜE, MARQUIS DE (1811–1892)
RECHERCHES THÉORIQUES ET EXPÉRIMENTALES SUR LES OSCILLATIONS DE L'EAU ET LES MACHINES HYDRAULIQUE À COLONNES LIQUIDES OSCILLANTES. Vol. 1, Versailles: E. Aubert (printer), 1880. Vol. 2, Paris: J. Baudry, 1883.
Vol. 1: [v] ii–vi, 384 pp.; 5 leaves of plates; 23 x 13 cm.
Vol. 2: [iv], 388 pp.; 3 leaves of plates (all folding); 23 x 13 cm.

A treatment of oscillatory flow in pipes, waves, pumps, fountains, locks and compressors.

388
LECHALAS, MÉDÉRIC CLÉMENT (1820–1904)
HYDRAULIQUE FLUVIALE. Paris: Baudry et c[ie], 1884.
[iv] 464 pp.; illus.; 23 x 14 cm. (Encyclopédie des Travaux Publics)

"Jules Charles & A. Brunet" on label pasted over publisher's imprint on title page.

Extensive treatise on river hydraulics by an inspector of the ponts et chaussées.

389
BOWIE, AUGUSTUS JESSE
A PRACTICAL TREATISE ON HYDRAULIC MINING IN CALIFORNIA. New York: D. Van Nostrand, 1885.
[v] 8–313 pp.; illus.; 17 leaves of plates (all folding); 25 x 15 cm. Microfilm copy.

390
SMITH, HAMILTON, JR. (1840–1900)
HYDRAULICS. The flow of water through orifices, over weirs, and through open conduits and pipes. London: Trubner and Co., 1886.
[v] vi–x [+i], 362 pp.; illus.; 20 leaves of plates (5 folding, 3 double); 30 x 24 cm. Presentation copy: "Clemens Herschel Esq. with the kindest regards and best wishes of Hamilton Smith, Jr. May 8th, 1886."

Known in 19th-century American hydraulics for his measurements of fluid flow, Smith inscribed this copy of the book to his contemporary, Clemens Herschel (see item 401).

391
BASSET, A. B. (1854–1930)
A TREATISE ON HYDRODYNAMICS. Cambridge [England]: Deighton, Bell and Co., 1888.
Vol. 1: [iii] iv–xii, 264 pp. [+16 pp. adv.]; illus.; 22 x 14 cm.
Vol. 2: [iii] iv–xv, [i errata], 328 pp. [20 pp. adv.]; illus.; 22 x 14 cm.

This work is similar to Horace Lamb's, whose influence is acknowledged by the author.

392
CHURCH, IRVING PORTER (1851–1931)
MECHANICS OF ENGINEERING. (Fluids) A treatise on hydraulics and pneumatics. New York: John Wiley & Sons, 1889.
[ii] iii–vii, 515–832, ix–xiii pp.; illus.; 24 x 13 cm. This volume forms part IV of the author's series: Mechanics of Engineering. Microfilm copy.

A very up-to-date work for its time, used as a textbook at Cornell University.

393
LILIENTHAL, OTTO (1848–1896)
DER VOGELFLUG, ALS GRUNDLAGE DER FLIEGEKUNST. Berlin: R. Gaertner, 1889.
[iii] iv–viii, 187 pp.; illus.; frontis. (col.); 8 leaves of plates (all folding); 23 x 15 cm.

394
ZHUKOVSKIĬ, NIKOLAĬ EGOROVICH (1847–1921)
SOBRANIE SOCHINENIĬ. Vol. 5: [Theoretical mechanics, hydrostatic and hydrodynamics, gravitational theory]. Moscow: State Publishing House of Technical and Theoretical Literature, 1949.
[7], 8–984 pp.; illus.; frontis.; 22 x 14 cm.

394a
ZHUKOVSKIĬ, NIKOLAĬ EGOROVICH (1847–1921)
POLNOE SOBRANIE SOCHINENIĬ. Vol. 1: General mechanics. Vol. 7: Hydraulics. Moscow: ONTI, 1937.
Vol. 1: [7] 8–636 pp.; illus.; 1 leaf of plates; 22 x 14 cm.
Vol. 7: [5] 6–410 pp.; illus.; 1 leaf of plates; 22 x 14 cm.

395
STRIZHEVSKIĬ, S.
NIKOLAI ZHUKOVSKY, FOUNDER OF AERONAUTICS. Translated by Y. Sdobnikov. Edited by H. Perham. Moscow. Foreign Languages Publishing House, 1957.
[iv], 92 pp. [1 p.] [+2 pp. adv.]; illus.; frontis.; 19 x 12 cm. (Men of Russian Science).

The first Russian hydrodynamicist of note, Zhukovsky demonstrated, among other things, that water-hammer pressures resulted from the propagation of elastic waves.

396
GANGUILLET, EMILE OSCAR (1818–1894) and WILHEM RUDOLPH KUTTER (1819–1888)
A GENERAL FORMULA FOR THE UNIFORM FLOW OF WATER IN RIVERS AND OTHER CHANNELS. Translated by Rudolf Hering and John C. Trautwine, Jr. Second edition. New York: John Wiley & Sons, 1901.

[v] vi–xxiii [i], 240 pp. [+16 pp. adv.]; illus.; 8 leaves of plates (all folding); 23 x 14 cm.

This work was apparently prompted by that of Humphreys and Abbot (see item 357) whom the Swiss authors criticised for not having included mountain torrents in their resistance formulation.

397
TAYLOR, DAVID WATSON (1864–1940)
RESISTANCE OF SHIPS AND SCREW PROPULSION. New York: The Macmillan Company, 1921.
[vi] vii–ix, 234 pp.; illus.; 1 leaf of plates (folding); 22 x 14 cm.

Taylor's work as a naval officer eventually led to the establishment of the great departmental test facility at Carderock, Maryland, that bears his name.

398
CREAVES, JOHN
A TREATISE ON ELEMENTARY HYDROSTATICS. Cambridge [England]: The University Press, 1894.
[v] vi–xi, 204 pp.; illus.; 18 x 12 cm.

399
HAZEN, ALLEN (1869–1930)
THE FILTRATION OF PUBLIC WATER-SUPPLIES. New York: John Wiley & Sons, 1895.
[iii] iv–x, 197 pp. [+2 pp. adv.]; illus.; 2 leaves of plates (1 double); 23 x 14 cm.

400
HERSCHEL, CLEMENS (1842–1930)
115 EXPERIMENTS ON THE CARRYING CAPACITY OF LARGE, RIVETED METAL CONDUITS. . . . New York: John Wiley & Sons, 1897.
[ii] iii–vi, 122 pp. [+16 pp. adv.]; illus.; frontis.; 13 leaves of plates; 23 x 14 cm.

A noted Yankee consulting engineer, Herschel wrote perceptively on hydraulic practice, developed and marketed what he called the venturimeter, and publicized the contributions of both Frontinus and Chézy.

401
KENT, WALTER G.
AN APPRECIATION OF TWO GREAT WORKERS IN HYDRAULICS: Giovanni Battista Venturi, born 1746; Clemens Herschel, born 1842. London: Blades, East & Blades, 1912.
[ii], 34 pp.; illus.; 25 x 19 cm.

402
BAZIN, HENRY ÉMILE (1829–1917)
EXPÉRIENCES SUR LA CONTRACTION DES VEINES LIQUIDES ET SUR LA DISTRIBUTION DES VITESSES DANS LEUR INTÉRIEUR. [Paris: Imprimerie Nationale, 1897]
[1] 2–63, [1] 2–27 pp.; illus.; 8 leaves of plates (all double); 26 x 19 cm. (Mémoires présentés par divers savants à l'Académie des Sciences de l'Institut national de France. t.23, No.4.) Bound with Expériences nouvelles sur la distribution des vitesses dans les tuyaux.

Originally a protégé of Henri Darcy, Bazin soon became an accomplished experimenter in his own right, especially on the characteristics of weir flow.

403
BAZIN, HENRY ÉMILE (1829–1917)
EXPÉRIENCES NOUVELLES SUR L'ÉCOULEMENT EN DÉVERSOIR EXECUTÉES A DIJON DE 1886 A 1895. Paris: Ch. Dunod, 1898.
[v] ii–vi, 200 pp.; illus.; 13 leaves of plates (12 folding); 24 x 16 cm. Résumé des articles parus dans les "Annales des ponts et chaussées" de 1888 à 1898.

A detailed study of flow over weirs of various cross-sectional shapes.

404
SULLIVAN, MARVIN E.
SULLIVAN'S NEW HYDRAULICS, consisting of new hydraulic formulas and the rational law of variation of coefficients. Denver: Mining Reporter Press, 1900.
[iii] 4–301 pp. [+7 pp. adv.]; illus.; 19 x 13 cm.

405
PRANDTL, LUDWIG (1875–1953)
GESAMMELTE ABHANDLUNGEN ZUR ANGEWANDTEN MECHANIK, HYDRO- UND AERODYNAMIK. Edited by Walter

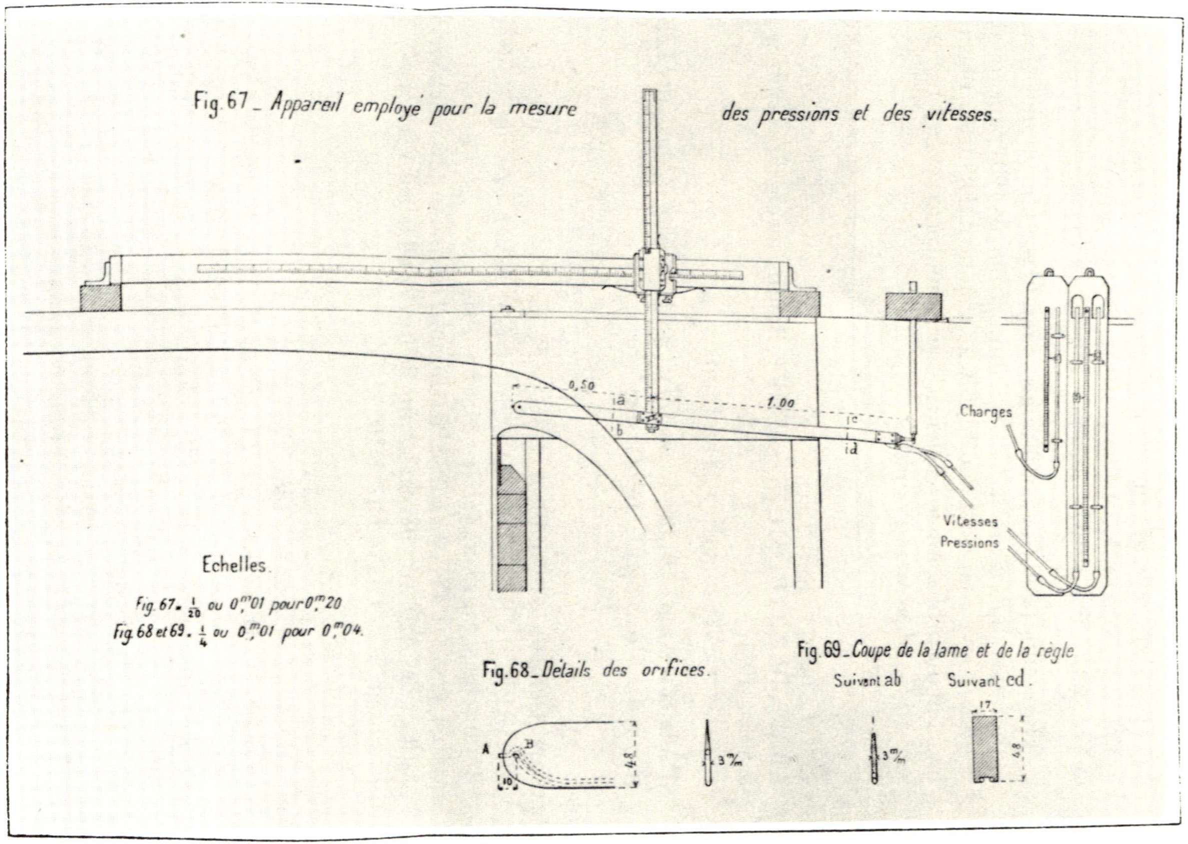

403. Apparatus used by Bazin in study of weir discharge.

Tollmien, Hermann Schlichting, Henry Gortler. Berlin: Springer-Verlag, 1961.
Vol. 1: [iv] v–xix, 574 pp.; illus.; frontis.; 23 x 15 cm.
Vol. 2: [v] vi–xiii, 575-1070 pp.; illus.; 23 x 15 cm.
Vol. 3: [v] vi–xiii, 1071–1620 pp.; illus.; 23 x 15 cm.

A long-time professor at Göttingen, Prandtl was known and beloved as the father of fluid mechanics, stemming from his 1904 paper on the boundary layer. His first and highly able student, Heinrich Blasius, later left the profession to teach in a boys' school because (typically) he said all his ideas were Prandtl's.

405a
PRANDTL, LUDWIG (1875–1953)
ABRISS DER STRÖMUNGSLEHRE. Braunschweig: Friedr. Vieweg & Sohn, 1931.
[iii] iv–vi, 223 pp.; illus.; 24 x 15 cm.

A later version of Prandtl's original (1913) booklet, "Abriss der Lehre von der Flüssigkeits- und Gasbewegung."

405b
PRANDTL, LUDWIG (1875–1953)
FÜHRER DURCH DIE STRÖMUNGSLEHRE. Zweite Auflage. Braunschweig: Friedr. Vieweg & Sohn, 1944.
[iii] iv–xi, [i], 384 pp.; illus.; 23 x 15 cm. Title page: "Zugleich vierte Auflage des Arbrisses der Strömungslehre des gleichen Verfassers."
Autograph: "L. Prandtl. 9. 10. 46."

406
VON KÁRMÁN, THEODORE (1881–1963)
COLLECTED WORKS OF THEODORE VON KÁRMÁN. Vol. 1, 1902–1913; Vol. 2, 1914–1932; Vol. 3, 1933–1939; Vol. 4, 1940–1951; [Vol. 5], 1952–1963. Vols. 1–4, London: Butterworths Scientific Publications, 1956. [Vol. 5] Rhode-St.-Genese, Belgium: Von Kármán Institute for Fluid Dynamics, 1975.
Vol. 1: [iv] v–ix, 531 pp.; illus.; frontis.; 25 x 15 cm.
Vol. 2: [iv] v–vi, 436 pp.; illus.; 25 x 15 cm.
Vol. 3: [iv] v–vi, 391 pp.; illus.; 25 x 15 cm.
Vol. 4: [iv] v–vi, 480 pp.; illus.; 25 x 15 cm.
[Vol. 5]: [iv] v–vii, 388 pp.; illus.; frontis.; 25 x 15 cm.

Von Kármán was undoubtedly the most noteworthy of Prandtl's

students, eventually surpassing even his mentor in productivity. A native of Hungary, he held professorships successively in Aachen and Pasadena, but lectured all over the world.

407
LANGLEY, SAMUEL PIERPONT (1834–1906)
EXPERIMENTS IN AERODYNAMICS. Second edition. City of Washington: The Smithsonian Institution, 1902.
[vii], 4–115 pp.; illus.; 17 leaves of plates (7 double) 33 x 24 cm. Smithsonian Contributions to Knowledge [Vol. 27].

An account of turntable and carriage tests on lifting vanes and propellers.

408
DUHEM, PIERRE (1861–1916)
RECHERCHES SUR L'HYDRODYNAMIQUE. Première série. Paris: Gauthier-Villars, 1903.
[iv], 211 pp.; 28 x 21 cm. "Extrait des Annales de la Faculté des Sciences de Toulouse, 2e Série, t. III, IV et V; 1901, 1902, et 1903."

Fundamental principles, propagation of discontinuities, of waves and of semi-waves.

409
DUHEM, PIERRE (1861–1916)
RECHERCHES SUR L'HYDRODYNAMIQUE. Deuxième série. Paris: Gauthier-Villars, 1904.
[iv] 153 pp.; 28 x 21 cm. "Extrait des Annales de la Faculté des Sciences de Toulouse, 2e Série, t. V; 1903."

Limiting conditions, theorem of Lagrange and the viscosity, coefficient of viscosity, and viscosity near the critical.

410
BELLASIS, E. S. (1855–1945)
HYDRAULICS, WITH WORKING TABLES. London: Rivingtons, 1903.
[v] vi–xii, 303 pp.; illus.; 22 x 14 cm.

A typical textbook by an executive engineer of the Indian department of public works.

411
MEAD, DANIEL W. (1862–1948)
NOTES ON HYDROLOGY AND THE APPLICATION OF ITS LAWS TO THE PROBLEMS OF HYDRAULIC ENGINEERING. [Chicago]: [Press of Shea Smith & Co.] 1904.
[ii] iii–xi, 202 pp.; illus. (some color); frontis.; 23 x 15 cm.

412
BUSQUET, RAYMOND (*b.*1855.)
PRÉCIS D'HYDRAULIQUE. LA HOUILLE BLANCHE. . . . Paris: J.-B. Baillière et Fils, 1905.
[v] vi–viii, 375 pp. [+ endpaper adv.]; illus.; 17 x 11 cm. (Encyclopédie industrielle)

A textbook used at l'École industrielle de Lyon.

413
WILLIAMS, GARDNER S. (1866–1931) and ALLEN HAZEN (1869–1930)
HYDRAULIC TABLES showing the loss of head due to the friction of water flowing in pipes, aqueducts, sewers, etc. and the discharge over weirs. New York: John Wiley & Sons, 1905.
[iv], 75 pp. [+16 pp. adv.]; illus.; 23 x 15 cm.

414
NEUMANN, FRITZ
DIE ZENTRIFUGALPUMPEN mit besonderer Berücksichtigung der Schaufelschnitte. Berlin: Julius Springer, 1906.
[v] vi–viii, 197 pp. [+2 pp. adv.]; illus.; 7 leaves of plates (all folding); 24 x 16 cm.

A well-illustrated treatment of pump design.

415
HOYT, JOHN CLAYTON (1874–1946) and NATHAN CLIFFORD GROVER (1868–1957)
RIVER DISCHARGE. Prepared for the use engineers and students. New York: John Wiley & Sons, 1907.
[ii] iii–vii [i], 137 pp.; illus.; 7 leaves of plates (2 folding); 23 x 15 cm.

A text on fluvial hydraulics by engineers of the U.S. Geological Survey.

416
EIFFEL, GUSTAVE (1832–1923)
RECHERCHES EXPÉRIMENTALES SUR LA RÉSISTANCE DE

L'AIR EXÉCUTÉES À LA TOUR EIFFEL. Paris: L. Maretheux, 1907.

[v] vi [i], 98 pp.; illus.; frontis.; 26 leaves of plates (9 folding, 5 double, 3 double-folding); 32 x 25 cm. Presentation copy: "à Raoul Dupain, souvenir affectueux de G. Eiffel."

Better known for the tower that bears his name, the French structural engineer Eiffel conducted many guided-fall and wind-tunnel tests on various aspects of human flight.

417
EIFFEL, GUSTAVE (1832–1923)

LA RÉSISTANCE DE L'AIR: examen des formules et des expériences. Paris: H. Dunod et E. Pinat, 1910.

[v] vi–vii, 258 pp. [+2 pp. adv.]; illus.; 1 folding plate); 22 x 14 cm.

418
EIFFEL, GUSTAVE (1832–1923)

LA RÉSISTANCE DE L'AIR ET L'AVIATION. Expériences effectuées au Laboratorie du Champ-de-Mars. Paris: H. Dunod et E. Pinat, 1910.

[v] ii–vii, 153 pp.; illus.; 57 leaves of plates (25 double, 3 double-folding, 1 laid in); 32 x 25 cm. Presentation copy: "A mon petit neveu Raoul Dupain, témoignage affectueux. G. Eiffel."

419
EIFFEL, GUSTAVE (1832–1923)

LA RÉSISTANCE DE L'AIR ET L'AVIATION. Expériences effectuées au Laboratoire du Champ-de-Mars. Deuxième édition, revue et augmentée. Paris: H. Dunod et E. Pinat, 1911.

[v] ii–vii, 252 pp.; illus.; frontis.; 59 leaves of plates (1 folding, 25 double, 3 double-folding, 1 laid in); 32 x 25 cm.

420
EIFFEL, GUSTAVE (1832–1923)

THE RESISTANCE OF THE AIR AND AVIATION. Experiments conducted at the Champs-de-Mars laboratory. Second edition, revised and enlarged. Translated by Jerome C. Hunsaker. London: Constable & Company Ltd., 1913.

[v] vi–xvi, 242 pp.; illus.; frontis.; 29 leaves of plates (all folding, one in back cover pocket); 31 x 25 cm.

417. Eiffel in the test section of his wind tunnel.

421
EIFFEL, GUSTAVE (1832–1923)
NOUVELLES RECHERCHES SUR LA RÉSISTANCE DE L'AIR ET L'AVIATION FAITES AU LABORATORIE D'AUTEUIL. Paris: H. Dunod et E. Pinat, 1914.
Vol. 1 (text): [v] ii–viii, 406 pp.; illus.; frontis.; 32 x 25 cm.
Vol. 2 (atlas): [vii] viii, 67 pp.; illus.; 72 leaves of plates (27 double, 6 double-folding); 32 x 25 cm.

422
EIFFEL, GUSTAVE (1832–1923)
RÉSUMÉ DES PRINCIPAUX TRAVAUX EXÉCUTÉS PENDANT LA GUERRE AU LABORATOIRE AERODYNAMIQUE EIFFEL, 1915–1918. Paris: Librairie Aéronautique, [1919].
[vii] viii–xx, 212 pp.; illus.; 7 leaves of plates (5 folding, one a duplicate); 31 x 24 cm. Two, duplicate, errata slips bound in.

423
UNWIN, WILLIAM CAWTHORNE (1838–1933)
A TREATISE ON HYDRAULICS. London: Adam and Charles Black, 1907.
[iv] v–xi [i errata], 327 pp.; illus.; 22 x 14 cm.

Oft-quoted textbook stemming from an encyclopedia article, by an emeritus professor of City and Guilds College, London.

424
UNWIN, WILLIAM CAWTHORNE (1838–1933)
A TREATISE ON HYDRAULICS. Second edition. London: Adam and Charles Black, 1912.
[iv] v–xi, 339 pp.; illus.; 22 x 13 cm.

425
FARGUE, LOUIS JÉROME (1827–1910)
HYDRAULIQUE FLUVIALE. La forme du lit des rivières à fond mobile. Paris: Gauthier-Villars, 1908.
[iii] 187 pp.; illus.; 30 leaves of plates (all double); 25 x 17 cm. (Encyclopédie des travaux publics)

Contains account of first river-model tests with movable bed.

426
GIBSON, ARNOLD HARTLEY (1878–1959)
HYDRAULICS AND ITS APPLICATIONS. London: Archibald Constable & Co., Ltd. 1908.
[v] vi–xiv [ii symbols, corrigenda], 757 pp.; illus.; 23 x 15 cm.

An assistant lecturer and eventually professor at Manchester, Gibson wrote what has been the most successful and durable English textbook on hydraulics and hydraulic machinery.

427
BOULANGER, AUGUSTE HENRI LÉON (1886–1923)
HYDRAULIQUE GÉNÉRALE. Vol. 1, Principes et problèmes fondamentaux; Vol. 2, Problèmes à signularités et applications. Paris: Octave Doin et Fils, 1909.
Vol. 1: [i] ii–xvi, 382, xii pp.; illus.; 18 x 12 cm.
Vol. 2: 1–299 pp.; illus.; 18 x 12 cm.; first signature missing.
(Encyclopédie scientifique. Bibliothèque de mécanique appliquée et génie.)

A textbook used at the École des ponts et chaussées.

428
BOVEY, HENRY TAYLOR (1852–1912)
A TREATISE ON HYDRAULICS. Second edition, rewritten. New York: John Wiley & Sons, 1909.
[ii] iii–xvii, 585 pp.; illus.; 23 x 15 cm.

An original treatment of hydraulics and hydraulic machinery by a Canadian professor at McGill.

429
FLAMANT, ALFRED AIME (*b.*1839.)
HYDRAULIQUE. [Third edition, revised and augmented] Paris: Librairie Polytechnique Ch. Béranger, 1909.
[iv] i–xvi, 699 pp. [+1 p. errata]; illus.; 24 x 15 cm. (Encyclopédie des travaux publics)

Flamant ranks with Gibson (English) and Forchheimer (German) for his comprehensive treatment of the subject.

430
TAYLOR, SIR GEOFFREY INGRAM (1886–1975)
THE SCIENTIFIC PAPERS OF SIR GEOFFREY INGRAM TAYLOR. Vol. 2, Meteorology, oceanography and turbulent flow. Vol. 3,

Aerodynamics and the mechanics of projectiles and explosions. Edited by G.K. Batchelor. Cambridge: Cambridge University Press, 1960.
Vol. 2: [vi] vii–x, 515 pp.; illus.; 2 pp., 1 leaf of plates; 26 x 18 cm.
Vol. 3: [vi] vii–xii, 559 pp.; illus.; 8 pp., 5 leaves of plates; 26 x 18 cm.

Taylor, a long-time fellow of Trinity College at Cambridge, ranks with Prandtl and von Kármán as one of the very great contributors to modern fluid mechanics, especially the theory of turbulence.

431
HUGHES, HECTOR JAMES (1871–1930) and ARTHUR T. SAFFORD (1867–1951)
A TREATISE ON HYDRAULICS. New York: The Macmillan Company, 1913.
[iv] v–xiv, 505 pp. [+6 pp. adv.]; illus.; 12 leaves of plates (5 folding); 23 x 15 cm.

432
BAKHMETEV, BORIS ALEKSANDROVICH (1880–1951)
O NERAVNOMERNOM DVIZHENII ZHIDKOSTI V OTKRYTOM RUSLE. St. Petersburg [Leningrad]: s.n., 1912.
[iii] iv–vii, 225 pp., 7 pp. of tables; illus.; 23 x 15 cm.

The author's St. Petersburg dissertation, still available in the Soviet Union, was translated and published in the United States as an Engineering Societies Monograph in 1912. The dissertation copy contains an English translation of the preface.

433
BAKHMETEV, BORIS ALEKSANDROVICH (1880–1951)
O NERAVNOMERNOM DVIZHENII ZHIDKOSTI V OTKRYTOM RUSLE. [With two supplementary papers: "K voprosu o raschete perepadov" and "O metode Prof. B. A. Bakhmeteva. . . ."] [A.A. Sabaneev, editor] Leningrad: Izdatel'stvo KUBUCH, 1928.
[iii] iv–xi, 275 pp.; illus.; 3 folding plates; 22 x 14 cm.

434
MACH, ERNST (1838–1916)
DIE MECHANIK IN IHRER ENTWICKLUNG HISTORISCH-KRITISCH DARGESTELLT. [Seventh revised and enlarged edition] Leipzig: F.A. Brockhaus; 1912.
[v] vi–xii, 494 pp.; illus.; 18 x 12 cm.

Mach's book is noted for its philosophical overtones.

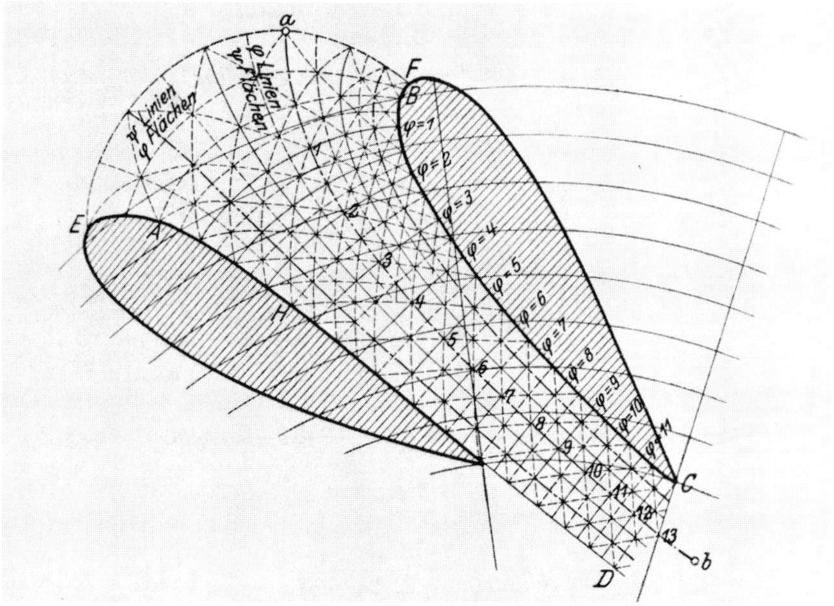

435. Early flow-net construction by Prášil.

435
PRÁŠIL, FRANZ (1857–1929)
TECHNISCHE HYDRODYNAMIK. Berlin: Julius Springer, 1913.
[iii] iv–viii, 269 pp. [+2 pp. adv.]; illus.; 23 x 16 cm.

A mathematical treatment with emphasis on the flow pattern.

436
PRÁŠIL, FRANZ (1857–1929)
TECHNISCHE HYDRODYNAMIK. [Second revised and enlarged edition] Berlin: Julius Springer, 1926.
[iii] iv–ix, 303 pp.; illus.; 23 x 15 cm.

437
GILBERT, GROVE KARL (1843–1918)
THE TRANSPORTATION OF DÉBRIS BY RUNNING WATER. Based on experiments made with the assistance of Edward Charles Murphy. Washington, [D.C.]: Government Printing Office, 1914.
263 pp.; illus.; frontis.; 2 leaves of plates; 29 x 22 cm. Department of the Interior, United State Geological Survey. Professional Paper 86.

A detailed account of what are probably the best-known experiments on sediment movement.

438
MARCHI, GIULIO DE (1890–1972)
SCRITTI DI GIULIO DE MARCHI. Milano: [Stabilimento grafico Tamburini] 1965.
[v], 363 [1] pp.; illus.; 2 leaves of plates (1 folding); 30 x 21 cm.

De Marchi was the Italian peer of Flamant, Gibson and Forchheimer, but a writer of papers rather than books.

439
VON MISES, RICHARD (1883–1953)
ELEMENTE DER TECHNISCHEN HYDROMECHANIK. I Teil. Leipzig und Berlin: B.G. Teubner, 1914.
[iii] iv–vii, 212 pp. [+4 pp. adv.]; illus.; 21 x 14 cm. Sammlung mathematisch-physikalischer Schriften, vol. 17, Part 1.

This little textbook was the first to reflect the modern fluid mechanics philosophy. Originally a German mechanical engineer, von Mises eventually became a noted Harvard mathematician.

440
Scobey, Frederick Charles (1880–1962)
THE FLOW OF WATER IN IRRIGATION CHANNELS. Washington, [D.C.]: Government Printing Office, 1915.
68 pp.; illus.; 20 pp. of plates; 23 x 14 cm. United States Department of Agriculture Bulletin No. 194. Professional Paper. May 10, 1915.

441
Scobey, Frederick Charles (1880–1962)
THE FLOW OF WATER IN WOOD-STAVE PIPE. With discussion by Gardner S. Williams, Theron A. Noble, D.C. Henny, E.A. Moritz, E.W. Schoder, L.M. Hoskins. Washington, [D.C.]: Government Printing Office, 1916.
96 pp.; illus.; 8 pp., 6 leaves of plates (5 folding); 23 x 14 cm. United States Department of Agriculture Bulletin No. 376. Professional Paper. November 25, 1916.

442
Scobey, Frederick Charles (1880–1962)
THE FLOW OF WATER IN CONCRETE PIPE. With discussion by Kenneth Allen, Arthur S. Bent, F.C. Finkle, Allen Hazen, J.B. Lippincott, and H.D. Newell. Washington, [D.C.]: Government Printing Office, 1920.
100 pp.; 8 pp., 4 leaves of plates (1 folding); 23 x 14 cm. United States Department of Agriculture Bulletin No. 852. Professional Paper. October 28, 1920.

443
Scobey, Frederick Charles (1880–1962)
COEFFICIENTS OF FLOW IN CONCRETE PIPE. [Baltimore] [s.n.], [1929].
30 pp.; illus.; 23 x 15 cm. Reprinted from the *Journal of the American Water Works Association,* Vol. 21, No. 1, January 1929.

444
Scobey, Frederick Charles (1880–1962)
THE FLOW OF WATER IN FLUMES. Washington, D.C.: U.S. Government Printing Office, 1933.
98 [+1] pp.; illus.; 16 pp., 2 leaves of plates (both folding); 23 x 14 cm. U.S. Department of Agriculture. Technical Bulletin No. 393. December, 1933.

445
RAMSER, CHARLES ERNEST (1885–1962)

THE FLOW OF WATER IN DREDGED DRAINAGE DITCHES. The results of experiments to determine the roughness coefficient, n, in Kutter's formula. Washington, D.C.: Government Printing Office, 1920.

60 pp.; 20 pp. of plates; 23 x 14 cm. United States Department of Agriculture Bulletin No. 832. Professional Paper. June 7, 1920.

446
YARNELL, DAVID LEROY (1886–1937) and SHERMAN M. WOODWARD (1871–1953)

THE FLOW OF WATER IN DRAIN TILE. Washington [D.C.]: Government Printing Office, 1920.

50 pp.; 6 pp., 7 leaves of plates (4 folding); 23 x 14 cm. United States Department of Agriculture Bulletin No. 854. Professional Paper. August 26, 1920.

447
EYDOUX, DENIS (*b.* 1876.)

HYDRAULIQUE GÉNÉRALE ET APPLIQUÉE. Paris: Librairie J.-B. Baillière et fils, 1921.

[v] vi–viii, 510 pp. [+2 pp. adv.]; illus.; 23 x 15 cm. Grandes encyclopédies industrielles J.-B. Baillière: Encyclopédie du génie civil et des travaux publics (vol. IX).

Like Boulanger (see item 427) Eydoux was a professor at the École des ponts et chaussées.

448
SPATARO, DONATO (1858–1928)

TRATTATO COMPLETO DI IDRAULICA TEORICA E SPERIMENTALE (le basi fisiche dell' idromeccanica). Vol. 1, Proprietà generali dei liquidi; Vol. 2, Equilibrio e moto dei liquidi; Vol. 3, Azioni—reazioni e resistenza dei fluidi. Milano: Ulrico Hoepli, 1924.

Vol. 1: [v] vi–xiv, 448 pp. [+1 p. errata]; illus.; 24 x 16 cm.
Vol. 2: [vii] viii–xii, 1151 pp. [+3 pp. errata]; illus.; 24 x 16 cm.
Vol. 3: [v] vi–xi, 985 pp. [+1 p. errata]; illus.; 24 x 16 cm.

Spataro was a professor at the Palermo school of engineering. His "Complete treatise" was indeed extensive: vol. I, general properties of liquids; II, equations of motion; III, action, reaction and resistance of fluids.

449
ALLIEVI, LORENZO (1856–1941)
THEORY OF WATER-HAMMER. Notes I to V. (Text). Translated by Mr. Eugene E. Halmos. Rome: Typography Riccardo Garroni, 1925.
[iii] iv–xi, 119 [+3] pp.; 28 x 19 cm.

This is a translation by E.E. Halmos of the 1902 paper by a prominent Italian engineer of Rome, "Teoria generale del moto perturbato dell'acqua nei tubi in pressione."

450
FLETTNER, ANTON (1885–1961)
THE STORY OF THE ROTOR . . . from the German "Mein Weg zum Rotor." [translated by F.O. Willhofft] New York: F.O. Willhoft, 1926.
[iv] 7–110 pp.; illus.; frontis.; 26 pp. of plates; 23 x 15 cm.

Propulsion of ships by wind force on rotating cylindrical stacks.

451
WESTCOTT, G.F.
HANDBOOK OF THE COLLECTIONS OF THE SCIENCE MUSEUM, SOUTH KENSINGTON, illustrating pumping machinery. Part I: Historical notes. Part II: Descriptive catalog. London: His Majesty's Stationery Office, Part I, 1932; Part II, 1933.
Part I: [ii], 3–103 pp.; cover illus.; 25 pp.; 4 leaves of plates (1 folding); 25 x 15 cm. Errata slip pasted in.
Part II: [ii], 3–194 [+1] pp.; cover illus.; 22 pp. of plates, 1 leaf of plates; 25 x 15 cm. Two slips pasted in: "Recent additions to the Pumping Machinery Collection" and "Corrections and Notes."
Handbooks of the Science Museum series.

452
FORCHHEIMER, PHILIPP (1852–1933)
HYDRAULIK. Dritte Auflage. Leipzig und Berlin: B.G. Teubner, 1930.
[iii] iv–x, 595 [1] pp. [4 pp. adv.]; illus.; 23 x 14 cm.

See remark under item 429.

453
ROUSE, HUNTER (*b.* 1906.) and SIMON INCE (*b.* 1921.)
HISTORY OF HYDRAULICS. Iowa City, [Iowa]: Iowa Institute of Hydraulic Research, State University of Iowa, 1957.

[vi] vii–xii, 269 pp.; illus.; 23 x 15 cm.

Stemming from a doctoral dissertation by the junior author, this greatly expanded work first appeared as a bilingual supplement to the journal "La Houille Blanche."

454
ROUSE, HUNTER (*b.* 1906.) and SIMON INCE (*b.* 1921.)
HISTORY OF HYDRAULICS. New York: Dover Publications, Inc., [1963].
[vi] vii–xii, 269 pp.; illus.; 22 x 14 cm.

455
ROUSE, HUNTER (*b.* 1906.) and SIMON INCE (*b.* 1921.)
HISTORY OF HYDRAULICS. Iowa City, Iowa: Iowa Institute of Hydraulic Research, The University of Iowa, [1980].
[vi] vii–xii, 269 pp.; illus.; 22 x 14 cm.

An unabridged and corrected republication of the work first published by the Iowa Institute of Hydraulic Research in 1957.

456
RIDJANOVIC, MUHAMED (1930–1983)
PRELIMINARY STUDY OF THE HISTORY OF HYDRAULICS IN RUSSIA FROM PREHISTORIC TIMES UNTIL THE 1917 REVOLUTION. Iowa City, [Iowa]: Mechanics and Hydraulics Department, College of Engineering, State University of Iowa, 1962.
[ii], 86 leaves; 1 folding plate; 28 x 21 cm. With inserted supplement, a review of the work, dated January, 1962, from the State Committee of Energetics and Electrification of the USSR, Federal Institute, "V.E. Vedeneev" for Scientific Investigation in Hydraulics. In Russian, with English translation.

Like the Bernoulli translations (items 101 and 123) and the early form of item 453, this was an Iowa dissertation, written for apparent lack of a comprehensive history of Russian hydraulics.

457
KOLUPAILA, STEPONAS (1892–1964)
BIBLIOGRAPHY OF HYDROMETRY. Notre Dame, Indiana: University of Notre Dame Press, 1961.
[iii] iv–xxiii, 975 pp.; frontis.; 23 x 15 cm.

The author, originally a Lithuanian professor, had an encyclopedic knowledge of hydrology, hydraulics and hydrometry and an extensive card index of some 22,000 items, twice lost during his migration following World War II.

458
ROUSE, HUNTER (*b.* 1906.)
HYDRAULICS IN THE UNITED STATES, 1776–1976. Iowa City, [Iowa]: Institute of Hydraulic Research, The University of Iowa, c1976.
[vi] vii–ix, 238 pp.; illus.; folding frontis.; 23 x 15 cm.

INDEX